# Limnology - The Importance of Monitoring and Correlations of Lentic and Lotic Waters

*Edited by Carmine Massarelli and Claudia Campanale*

Published in London, United Kingdom

Limnology - The Importance of Monitoring and Correlations of Lentic and Lotic Waters
http://dx.doi.org/10.5772/intechopen.100936
Edited by Carmine Massarelli and Claudia Campanale

Contributors
Ewelina Widelska, Barbara Sowińska-Świerkosz, Wojciech Walczak, Sara C. C. Antunes, Sara Rodrigues, Ivo Pinto, Sandra Nogueira, Harsh Gulati, Sarita Rana, Luis M. Mejia-Ortiz, Oscar Frausto-Martínez, Alejandro L. Collantes-Chávez-Costa, Cruz López-Contreras, Carmine Massarelli, Claudia Campanale, Vito Felice Felice Uricchio, Nqobizitha Siziba, Emmanuel Tapiwa Sero, Sharangouda J. J. Patil, N. Eramma, C.N. Venkatesh, H.M Lalitha, S. Satishgouda, S. Renuka Jyothi



First published in London, United Kingdom, 2023 by IntechOpen
IntechOpen is the global imprint of INTECHOPEN LIMITED, registered in England and Wales, registration number: 11086078, 5 Princes Gate Court, London, SW7 2QJ, United Kingdom

British Library Cataloguing-in-Publication Data
A catalogue record for this book is available from the British Library

Additional hard and PDF copies can be obtained from orders@intechopen.com

Limnology - The Importance of Monitoring and Correlations of Lentic and Lotic Waters
Edited by Carmine Massarelli and Claudia Campanale
p. cm.

This title is part of the Environmental Sciences Book Series, Volume 2
Topic: Ecosystems and Biodiversity
Series Editor: J. Kevin Summers
Topic Editors: Salustiano Mato, Josefina Garrido and Francisco Ramil

Print ISBN 978-1-80356-536-1
Online ISBN 978-1-80356-537-8
eBook (PDF) ISBN 978-1-80356-538-5
ISSN 2754-6713

# IntechOpen Book Series
# Environmental Sciences
## Volume 2

## Aims and Scope of the Series

Scientists have long researched to understand the environment and man's place in it. The search for this knowledge grows in importance as rapid increases in population and economic development intensify humans' stresses on ecosystems. Fortunately, rapid increases in multiple scientific areas are advancing our understanding of environmental sciences. Breakthroughs in computing, molecular biology, ecology, and sustainability science are enhancing our ability to utilize environmental sciences to address real-world problems.

The four topics of this book series - Pollution; Environmental Resilience and Management; Ecosystems and Biodiversity; and Water Science - will address important areas of advancement in the environmental sciences. They will represent an excellent initial grouping of published works on these critical topics.

# Meet the Series Editor

J. Kevin Summers is a Senior Research Ecologist at the Environmental Protection Agency's (EPA) Gulf Ecosystem Measurement and Modeling Division. He is currently working with colleagues in the Sustainable and Healthy Communities Program to develop an index of community resilience to natural hazards, an index of human well-being that can be linked to changes in the ecosystem, social and economic services, and a community sustainability tool for communities with populations under 40,000. He leads research efforts for indicator and indices development. Dr. Summers is a systems ecologist and began his career at the EPA in 1989 and has worked in various programs and capacities. This includes leading the National Coastal Assessment in collaboration with the Office of Water which culminated in the award-winning National Coastal Condition Report series (four volumes between 2001 and 2012), and which integrates water quality, sediment quality, habitat, and biological data to assess the ecosystem condition of the United States estuaries. He was acting National Program Director for Ecology for the EPA between 2004 and 2006. He has authored approximately 150 peer-reviewed journal articles, book chapters, and reports and has received many awards for technical accomplishments from the EPA and from outside of the agency. Dr. Summers holds a BA in Zoology and Psychology, an MA in Ecology, and Ph.D. in Systems Ecology/Biology.

# Meet the Volume Editors

Carmine Masserelli is an environmental technologist at the Water Research Institute (IRSA) of the Italian National Research Council (CNR). He is an expert in the development of smart technologies for water management and environmental monitoring, characterisation and monitoring of contaminated and degraded sites, integration of spatial data such as standard methodologies, interoperability, and spectral data infrastructures. He also has expertise in the design of complex monitoring programs of water and soil aimed at the correct assessment of the environmental status. Dr. Masserelli is currently coordinating several research projects for regional monitoring of pesticide residues in water bodies, pollution by nitrates of agricultural origin, understanding the dynamics of environmental illegals, and safeguarding biodiversity.

Claudia Campanale is a senior post-doc research fellow at the Water Research Institute (IRSA) of the Italian National Research Council (CNR). She obtained a Ph.D. in Chemical and Molecular Sciences from the Department of Chemistry, University of Bari, Italy. Her research interests include specific awareness, knowledge, and familiarity with emerging pollutants monitoring (e.g., pesticides, PCBs, microplastics, heavy metals) in environmental matrices. Dr. Campanale has consolidated experience in chemical analytical techniques such as HPLC MS/MS, GC-MS/MS, and ICP-MS to characterise and quantify pollutants (inorganic and organic) in environmental matrices (waters, soils, vegetables, sediments).

# Contents

# Preface

Monitoring the activities of water bodies is an effective way to understand the state of the aquatic environment and provides important information for territorial planning.

New environmental regulations require monitoring activities in water bodies to establish the state of environmental quality of each of them. The knowledge of the state of the water bodies allows for their classification and, if necessary, to plan their rehabilitation to achieve environmental quality objectives. All water bodies that have a significant environmental interest and those that may have a negative influence due to being highly polluted should be monitored.

This book examines methods for implementing effective monitoring of water resources and all aquatic environments in general. It discusses how to monitor the quality of the water matrix through ecotoxicological tests and indicator species. It highlights the importance of monitoring and its correct management for long-term protection. It also proposes methodological updates for more effective monitoring, aimed at acquiring more information in detail useful to territorial administrators who are often forced to make decisions without being supported by scientific evidence. Finally, the book presents an updated review of emerging contaminants for these critical environments that need more attention.

**Carmine Massarelli and Claudia Campanale**
National Research Council,
Water Research Institute,
Bari, Italy

Section 1

# Correlation Aspects

Chapter 1

# Zooplankton Productivity Evaluation of Lentic and Lotic Ecosystem

*N. Eramma, H.M. Lalita, S. Satishgouda, S. Renuka Jyothi, C.N. Venkatesh and Sharangouda J. Patil*

## Abstract

The present study reveals the correlation of zooplankton productivity of lotic and lentic water ecosystem. The biological rhythms are useful tool to determine the zooplankton production. Planktons enrich the trophic level of lentic and lotic ecosystems. The relationship of primary productivity of any aquatic ecosystem depends on the biological biodiversity. Estimation of zooloplankton productivity of any water system helps to analyse its richness of species as biological population which are sustainable in it by the adequate amount physicochemical parameters. It is observed in most of the water system the quality of physical, chemical and biological phenomena are depends directly or indirectly correlation to establish diverse life as suitable habitat. To identify the current problems on lentic and lotic water system correlation studies will be more helpful, it is need of hour to give more attention on limnology because most of the biodiversity polluted due to industrialisation and anthropology. To develop advanced monitoring tools to address it on time to time problems of water system as key indicators and to conservation strategies towards the sustainable developments.

**Keywords:** ecological adaptation, climate change, conservation, river basin, physicochemical, heavy metal, sediments, microorganisms, water quality, water pollution, anthropology, diversity

## 1. Introduction

Water is one of the most significant resources on earth and is fundamental to all kinds of life since it is necessary for the survival and well-being of all living things [1]. The presence of planktonic organisms is the distinctive peculiarity of all aquatic ecosystems, whether they are lentic or lotic. Planktons are typically regarded as a measure of the water column's fertility [2]. Studies on fresh water sources, whether they are man-made or natural, have become increasingly important in recent years, mostly due to their variety of uses.

In aquatic system, surface water called as inland water ecosystem and classified into lotic and lentic system. The land surface habitats for free water and will be categorised as lotic or lentic system. In the inland water system, act as fundamental habitat as ecological constraints to divide in to as lotic and lentic system [3]. Lentic water ecosystem exhibits discrete habitat as aquatic matrix in the terrestrial system, whereas in lotic water ecosystem shared continuous habitat with the linkage of various basins in unidirectional flow in the dendritic structure of river networks. These ecosystems act considerably different in physical, chemical and biological characteristics (**Table 1**). Lotic water ecosystem significantly differentiated when compare to lentic water ecosystem, which possess relative affinity with terrestrial waters. In the lotic water ecosystems, waters flow establish it continuous, definite and unidirectional approach in the form of measurable, constant flow, wherein lentic water ecosystems, waters not exit in any flow in the continuous, definite and any direction in the given medium as staging structure. In lotic water ecosystem, flow of water source towards flat surface, whereas in lentic system water equally stored as per the specific landscape, terrestrial system and topography of the diversity. The lotic water system walks their path in the narrow form, shallow level, broader surface, relatively rapid, slow moving and deeper in nature. Lotic water system will be diverse and area specific in their structure, reaching small area as a spring in a centimetre wide exhibit like a major river which covers kilometre width. These water systems showed main difference in the communities, they move as per the velocity and along with lotic communities. These systems depends which are the organisms occupied the ecosystem and richness of diversity exist in the ecological niche. In this context, the plant diversity in lotic community is lower compared to community of lentic water system while small components of lotic community exhibit similar environment in the lentic community. These plants have adaptive evolutionary significance to cope up with the environmental force and various conditions that brings by running water. Such biological adaptations have placed different type of species successfully to take advantage in the lotic water system as one of the ecological niche. In this aquatic ecosystem, water differs in their stability, persistence basically, determining the structure of population and their genetics, size of geographical range and rate of speciation in lentic and lotic lineages [4]. Whereas the pattern of this phenomenon is remain incomplete and not understood, the compatibility of such traits preparedness with all the ability of species to develop as new populations to bring it as significant differences in the system [4]. Due to water current velocity, affect the processing time directly, in indirect approach leaf fragmentation is common to renew the dissolved oxygen and supporting for microbial diversity. The contrasting feature of geological activities and ecological strength of these water systems exhibited relative index of evolutionary and biogeographical differences in the species in different habitats [5]. Species taxa is depends on geological period and they were ecologically unstable due to adoptive condition to lentic habitat and found less population and greater area of geography when compare to the range of lotic relative species, such pattern observed through the range of selected taxa, specific regions and ecosystem [6, 7].

Ecosystem of lentic and lotic significantly differ basically in the local environmental conditions (For example, presence or absence of flow, time of water residence) and property of physical connectivity. While lotic water system not suffer much from fragmentation and deviation than the lentic water system, climate-induced changes from permanent to temporary locales will reduce the connectivity in the equal landscape ecosystem in in lotic system. In these environment exhibits harsh conditions for specific plant species, in those larger plants more prone, herbaceous species adjust with such conditions and acts as more favourable for the lotic community. It was noted that they were more tend to available physical conditions to adopt such water. Algal species can adapt to all sorts of places, surfaces and hence, they take significance as successful feature of the lotic water system. All the algal species have developed genetic ability with evolutionary changes to adapt over the period of their times to

| Sl. No. | Characteristics | Lotic system | Lentic system | References |
|---|---|---|---|---|
| 1 | Water flow | Water flows in a continuous and definite direction (flowing waters/running waters/moving waters) | Absent; water do not flow in a continuous and definite direction (standing waters/static waters/still waters/ stationary waters) | [3–8] |
| 2 | Morphometry of water body | Linear, longer, narrower and shallower basin with more complex perimeter | Circular and deeper basin with less complex perimeter | |
| 3 | Width of water bodies | Relatively narrow | Relatively broad | |
| 4 | Depth of water | Shallower | Deeper | |
| 5 | Landscape position | Natural placement | Lower in the catchment | |
| 6 | Regional distribution and function | Dictated by geomorphology | Influenced by regional demand for dam related ecosystem services | |
| 7 | Permanence | Can last many thousands of years | Usually only last for a few hundred to a few thousand years | |
| 8 | Source of water | Fed by lower order streams and diffuse inputs; fed by rains, precipitation, snow melt and springs | Fed by higher order streams, surface water dominated; in addition to ground water, lotic waters from rivers, streams and creeks drain into lakes and ponds to form a lentic environment | |
| 9 | Current velocity | High | Low | |
| 10 | Water retention time/ water residence time (how long it holds water) | Sometimes there is a lot of water like after a heavy rain; sometimes there is a very little water during drought; will dry up and many organisms will diev | Last longer and organisms can continue to live despite the shortened supplies | |
| 11 | Dissolved oxygen content | Usually high in dissolved oxygen; higher percentage of dissolved oxygen in water, due to water flowing | Lower percentage of dissolved oxygen, especially in deeper water | |
| 12 | Salt content | Lower salt content | Higher salt content as water evaporates over time | |
| 13 | Adaptability of creatures | The creatures that live here must adapt to their past-paced environment | The creatures that live here must also adapt to their environment | |
| 14 | Speciation rates | Low | High | |
| 15 | Geographical range size | Higher | Lower | |
| 16 | Species diversity | Higher | Lower | |
| 17 | Stability | High | Low | |

**Table 1.**
*Characteristics of lotic and lentic aquatic systems.*

prevent and overcome against water current from sweeping it completely. Dry season disturb the fauna and flora which occurs in both the aqua system globally and it can be strongly affect the communities. Whereas the drying effect depends on ecosystem and their relationship with local species ability strengthens the community, dryness is completely different with climate and their community responses between lentic and lotic water system comparatively [8].

Researchers predicted that drying of fauna and flora in aquatic communities may effect strongly in isolated lentic water system compare to scattered lotic system, if provided support of hydrological connectivity in higher level reduce the drying effect passively rather than dispersers, because such sources may potentially greater in recolonizing ability of the species in the ecosystem. Whereas such variation, strongly combine the quality characteristics to make the healthy habitat of running waters compare to still waters. Flow of water should be unidirectional also observed that spatial and temporal heterogeneity in all the scales (microhabitats) in potent level and high degree [8].

There are two sorts of natural freshwater bodies: lotic and lentic. Running water is referred to as lotic since the entire body of water moves in that direction. Brooks, streams, rivers, and springs, which are representative of the lotic bodies in India, may be included in these [9]. Lentic ecosystems such as pools, ponds, swamps, bogs, lakes, exhibit a wide range of chemical, physical and biological characteristics (**Figures 1–3**). In general, they have different zonation, unlike the rapid and the pool zones of the lotic system with different specialised lentic community.

Limnology is the study of all aquatic systems, both lentic and lotic fresh, fresh, and saline including lakes, wetlands, marshes, bogs, ponds, reservoirs, streams, rivers, oceans, etc. about their physical, chemical, and biological characteristics [10].

The physical, chemical, and biological properties of lakes vary greatly. They differ physically in terms of temperature, water circulation, and light intensity. They differ biologically in terms of biomass, population size, and growth as well as chemically in terms of nutrients, major ions, and pollutants [11]. Management of freshwater bodies should aim to maintain high productivity level of water bodies with provision for a

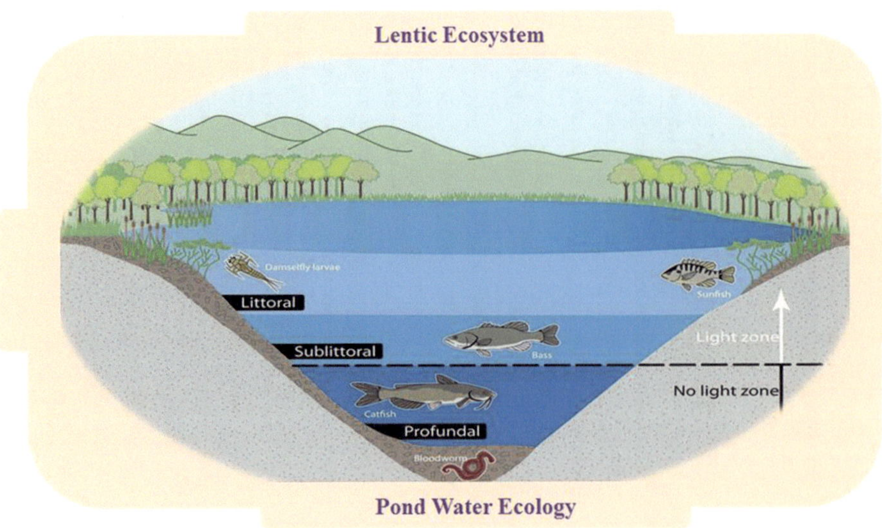


**Figure 1.**
*Types of lentic system and their morphology.*

**Figure 2.**
*Anatomy of the aquatic ecosystem represents the zones and species diversity.*

high rate of harvest of plants and animals for human use. The amount of solar energy trapped by the autotrophic organisms is known as production. The amount of solar energy trapped by the autotrophic organisms in unit time is known as productivity. The amount of plankton present in the water body decides the productivity of that water body [12]. Primary productivity is inversely related to the nutrient concentration, indicating that when the nutrient status is low, both primary productivity and fish output decline. There is only one primary production peak in reservoirs during the summer or early summer, according to numerous researchers who have examined the primary productivity of water bodies in different places and at different times of the year [10].

Zooplanktons are a crucial component of the aquatic ecology. Zooplanktons are essential components of the food chain and biogeochemical processes. They serve as helpful indicators for environmental changes [13]. Water currents carry zooplankton from lentic to lotic systems, and the energy in their tissues can affect the amount of resources available downstream, influencing ecological functions and the community structure of lotic consumers [14]. Zooplanktons are influenced by a wide range of environmental variables, such as pH, temperature, salinity, oxygen, and others [13].

## 2. The lentic ecosystems

The term lentic (meaning 'to make calm') is used for still waters of lakes and ponds, which offer environmental conditions, which differ sharply with that of the streams [15, 16]. Lentic aquatic systems contain stagnant waters. These are formed

**Figure 3.**
*Types of lotic system and their morphology.*

usually in large or small depressions formed on earth's surface where water is collected and cannot flow out. These water bodies include ponds, lakes, swamps etc. As a result, the lentic systems are closed systems; most aquatic life rarely leaves them after entering. Within the lentic body, they must endure, deteriorate, and decompose. They eventually transform into swamps or marshy wetlands before becoming dry lands with time. The depth to which light can penetrate depends on the turbidity. The temperature varies with depth and the season. The oxygen level of the lentic ecosystem is considerably lower than that of the lotic because only a small portion is in direct contact with the atmosphere and because decomposition proceeds intensively at the bottom.

### 2.1 Zonation in lentic systems

The bodies like ponds and lakes have three zones (**Figure 3**):

- Littoral Zone
- Limnetic Zone
- Profundal Zone

#### *2.1.1 Littoral zone*

The area of shallow water is where light can reach the ground. This area is often identified by the presence of rooted plants because it receives a lot of sunlight. This might not apply to "managed" or "man-made" lakes or ponds.

#### *2.1.2 Limnetic zone*

The compensation level is the area of open water up to the depth of effective light penetration. The depth at which the rate of respiration and photosynthesis are equal is known as the compensation level. This depth will typically be at the point when light intensity is 1% of that of full sunlight. Only planktons, nektons, and occasionally neustons make up the community in this zone. Small, shallow ponds lack this zone. The term "Eutrophic Zone" refers to the entire lighted stratum, including the littoral and limnetic zone.

#### *2.1.3 Profundal zone*

This is the deep bottom region, past the point at which light can effectively penetrate. Ponds frequently lack this zone.

## 3. The lotic ecosystem

Running water is referred to be lotic (from the Greek word lavo, which means "to wash"), in which the entire body of water moves in one direction. These could include springs, rivers, streams, and lakes. A tiny natural body of water is referred to as a brook, whereas a comparatively big natural body of water is referred to as a river. The stream is typically described as being larger than a brook but smaller than a river. Water from the earth, which appears on the surface as a stream in the spring, is a problem [15, 16]. The current's speed, which can produce either slow-moving or fast-moving streams with highly different features, essentially determines the lotic [17]. Plant and animal communities in slow-moving streams typically resemble those in lentic (lake and pond) habitat. According to Tokeshi and Pinder [18], algae and plants are crucial to lotic systems as a source of energy, for building microhabitats that protect other animals from predators and the current, and as a food source [18, 19].

## 4. Physico-chemical characteristics of lentic ecosystems

Closed systems are lentic systems. Due to the lack of an exit for the water body, persistent substances, such as the byproducts of the decomposition and mineralisation of organic matter and the degraded or partially degraded pollutants that are released into these aquatic bodies, continue to exist in the system. The biotic communities are significantly impacted by this. Due to the lack of water movement, the physico-chemical environment of lentic aquatic systems is exceptional. The following list of key characteristics of lentic water bodies includes:

### 4.1 Water quality

Physical–chemical study of a water sample provides a picture of the physical and chemical components, however this analysis simply provides a numerical value. To determine the precise quality of water, an indexing method called the "Water Quality Index (WQI)" has been devised. WQI provides information on the whole aquatic system's quality. "A rating of water quality, which shows the composite influence of different water quality criteria on the overall quality of water," is how the term "water quality index" is defined [20]. Physical, chemical, and biological characteristics of water must all be considered when evaluating its quality because they are interconnected [21]. The majority of the water in the lentic system comes from rainfall, surface runoff, or subsurface water sources. The makeup of these sources does not alter all that much over time. As a result, the water quality in lentic systems has essentially not changed over a lengthy period of time.

### 4.2 Seasonal changes

A lentic body's positive association with seasonal variations in the physicochemical properties of water affect how productive it is. The make-up of the living community is significantly impacted by the seasonal variations in water quality. A significant increase in phytoplankton is seen during periods of intense sunlight, which is also accompanied by a striking decrease in the concentration of numerous plant nutrients.

### 4.3 Assessment of productivity

The entire surface area of the water body is more significant than the water volume or the depth of the water body when determining the productivity of a lentic water body. The productivity is based on the combined length of the limnetic and littoral zones. Even if some lentic bodies' profundal zones are nutrient-rich, productivity is scarcely impacted by a lack of sunlight.

### 4.4 Stratification and water movement

One of the most significant characteristics of a lentic water body is the stratification phenomena, often known as vertical zonation. The variation in density brought on by the varied heating of lake waters causes stratification to exist. If the temperature stays consistent at more than $4.^{0}$ C when there are significant breezes present, the lake water is well mixed. This is seen in lakes and ponds where the static waters are often deeper, measuring more than 6 to 8 metres. Because of stratification, the lentic water's many strata range in temperature, oxygen concentration, and nutrient status [22]. The lake is stratified into the epilimnion, hypolimnion, and thermocline when the temperature is not constant due to density differences. Lakes are categorised into amictic, meromictic, holomictic, oligomictic, monomictic, dimictic, and polymictic lakes based on their circulation patterns. Thus, the temperature and wind pattern have a significant impact on how the water moves. Water in lakes frequently moves in multiple directions.

### 4.5 Effect of light

The depth to which light can penetrate depends on the turbidity. The temperature varies with depth and the season. The spectral makeup and intensity of the light there govern how deep rooted macrophytes and associated algae can develop on suitable surfaces.

A lake can be separated into tropogenic zone and tropolytic zone based on the amount of light it receives. The primary production of a lake is influenced by light, and phytoplankton in turn affects the depth of light penetration. The oxygen level of the lentic ecosystem is considerably lower than that of the lotic because only a tiny fraction is in direct contact with the atmosphere and because decomposition occurs intensively at the bottom.

## 5. Physico-chemical characteristics of lotic ecosystems

### 5.1 Currents and stream pattern

Running waters' gradient and substrates have an impact on the current's speed. Wind has little effect on currents in running waterways, in contrast to lentic waters. The properties of the drainage basin are what determine how water, dissolved chemicals, and suspended particles travel continuously downstream. According to this gradient, there are numerous stream patterns, including dendritic, rectangular, radial, trellised, parallel, annular, deranged, and pinnate. The risks of soil erosion are determined by the stream pattern.

### 5.2 Light

Light is important to lotic systems, because it provides the energy necessary to drive primary production via photosynthesis, and can also provide refuge for prey species in shadows it casts. The amount of light that a system receives can be related to a combination of internal and external stream variables [23]. The presence of turbidity has a significant impact on how well light penetrates moving waters. There is a loss due to water absorption in addition to scattering by particles. A sufficient amount of light can reach the substrate and enable photosynthesis if the water is transparent or hollow. The availability of light may also be affected by seasonal and nocturnal circumstances because the angle of incidence—or the angle at which light strikes water—can cause light to be reflected back into space [24].

### 5.3 Temperature

Temperature is a crucial abiotic component for most lotic species since they are poikilotherms, whose internal temperature fluctuates with their surroundings. Water can be heated or cooled through surface radiation, conduction to or from the air, and other nearby substrates [24]. The absence of temperature-related stratification means that a stream's temperature follows that of the surrounding air because of increased interaction with it. Numerous variables, including as the source, depth, substrate, tributaries, exposure, and time of day, affect the temperature of lotic water. The contribution of surface and ground waters to stream flow varies depending on a number of variables, including the geology and temperature of the area. Running water fuelled mostly by surface runoff has variable flow and may spate after each significant downpour, whereas running water fed primarily by ground water often has regular flow.

### 5.4 Dissolved gases

The most prevalent and significant dissolved gas in flowing water is oxygen. Due to turbulence and mixing, there is a high concentration of oxygen. Organic contamination is typically indicated by low concentration. However, the oxygen concentration in the diurnal basis varies. Current, water temperature, and the presence of breathing plants and animals all affect how much oxygen is present. The carbon-di-oxide concentration of the moving waters tend to be sparse due to constant turbulence of water and its frequent interaction with air [3].

## 6. Distribution of planktonic forms in lentic and lotic systems

Plankton distribution patterns and environmental conditions are closely connected. Climate, water, temperature, light intensity, nutrient concentration, river shape, discharge, water residence time, precipitation, and biotic variables are examples of potential physical, chemical, and hydrological elements [25]. All the creatures suspended in unrestricted water are referred to as plankton. Aquatic organisms that drift passively and have limited ability to move in opposition to the movement of the water mass make up the plankton. Compared to benthic or nektonic species, planktonic organisms have a short life cycle and high metabolic activity, which allows them to respond to any pollution challenge quickly and dramatically [26]. Hence, study of planktonic community is of crucial importance in understanding pelagic productivity and pollution impacts [2].

Phytoplankton and zooplankton are two types of plankton. All suspended microalgae in a body of water that belong to all taxonomic algal groups are referred to be phytoplankton [27]. The primary producers in aquatic habitats and the foundation of the food chain are phytoplankton and other aquatic plant life [28]. Zooplankton, the animal element of the plankton, is a vital component of a freshwater ecosystem's food chain since it occupy a central position between of autotrophs and other heterotrophs. All varieties of aquatic bodies exhibit zooplankton abundance due to energy transfer occurring at various trophic levels [9].

Pelagic creatures known as zooplankton are those that are unable to hold their location by swimming against the physical movement of water [29]. They were found in practically all water bodies, including rivers, streams, lakes, reservoirs, ponds, irrigation canals, rice fields, and temporary water bodies. They lived in both freshwater and saltwater. By regulating phytoplankton production and modifying the pelagic ecology, zooplankton play a crucial role in the pelagic food web. They are heterotrophic animals unable to produce organic materials on their own [30].

Water currents transport zooplankton from lentic to lotic systems, and the energy in their tissues can change resource availability downstream, changing ecosystem functioning and the community structure of lotic consumers [31]. As essential biotic components of the food webs in any body of water, zooplankton diversity and abundance play a crucial role in the establishment of water quality and trophic levels as well as serving as the subject of bioindication and environmental condition monitoring [32]. The zooplankton community is quite susceptible to changes in the environment. Therefore, they are of ecological relevance because changes in their abundance, species diversity, or community composition can give key signs of environmental change or disturbance [10].

## 7. Primary productivity of lentic and lotic ecosystem

Any aquatic ecosystem's main production is based on the diversity of planktonic organisms. A water body's primary productivity estimate can be used to determine how much biological population it can support through respiration. It is the most significant biological phenomenon, on which all forms of life are directly or indirectly dependent [11]. The amount of primary production must be determined in order to gauge a reservoir's bioactivity. Production is the measure of how much solar energy is captured by autotrophic organisms. Productivity is the solar energy captured by autotrophic organisms per unit of time [33]. The rate at which radiant energy is stored by a producer's photosynthetic and chemosynthetic processes is referred to as primary productivity [34]. The most significant biological phenomenon in nature is primary productivity, which is directly or indirectly dependent on the entire biodiversity. Through the process of photosynthesis, primary producers generate organic matter from inorganic nutrients. Primary producers need essential nutrients to live and grow such as nitrogen, phosphorus, magnesium, calcium, iron, zinc, etc. in sufficient amount. Phytoplankton, Macrophytes, and Periphyton are the main producer in lake and reservoir [35].

Primary producers require appropriate amounts of key nutrients including nitrogen, phosphorus, magnesium, calcium, iron, and zinc to survive and flourish. The primary producers in lakes and reservoirs include phytoplankton, macrophytes, and periphyton [36]. The balance between gross photosynthesis, respiration, and other plant losses, such as death, is referred to as net primary productivity. There are several characteristics of primary production in aquatic ecosystems that are different from those in terrestrial systems. Contrary to nutrients, which may go through multiple cycles, the movement of energy through an ecosystem is a one-way process. The primary productivity is the foundation of all food chains and food webs in each ecosystem, and it also produces 70% of the world's atmospheric oxygen. Note that the efficiency with which energy is converted by organisms into other forms (chemical energy) and the quality of incident solar energy per unit area of the ecosystem are the two main concerns of ecologists interested in ecoenergetics and the study of productivity, which is currently receiving much attention in ecology [37].

The productivity of a water body is determined by the quantity of plankton present within. When compared to lotic water, lentic water has the highest primary productivity. This suggests that lentic water has more planktonic activity. The "light and dark bottle method" was used to determine the primary organic output of the river water. The method of measuring the production and consumption of oxygen using light and dark bottles and Winkler's titration was initially put forth by Gaarder and Gran in 1927 [38]. It is the most typical technique for assessing output and productivity in the aquatic environment. The amount of dissolved (or free) oxygen in water or wastewater is measured. The amount of dissolved gas (oxygen) per litre of water is known as the dissolved oxygen concentration. In this method, productivity is calculated by evaluating the dissolved oxygen content (**Figure 4**).

## 8. Water productivity calculation

The following formulas were used to compute Gross Primary Productivity (GPP), Net Primary Productivity (NPP), and Respiration:

$$Gross\ Primary\ Productvity = \frac{LB - DB}{\mathrm{T}} \mathrm{X} \frac{0.375}{PQ} \mathrm{X}\, 1000mg\,/\,\mathrm{L}\,/\,\mathrm{h}$$

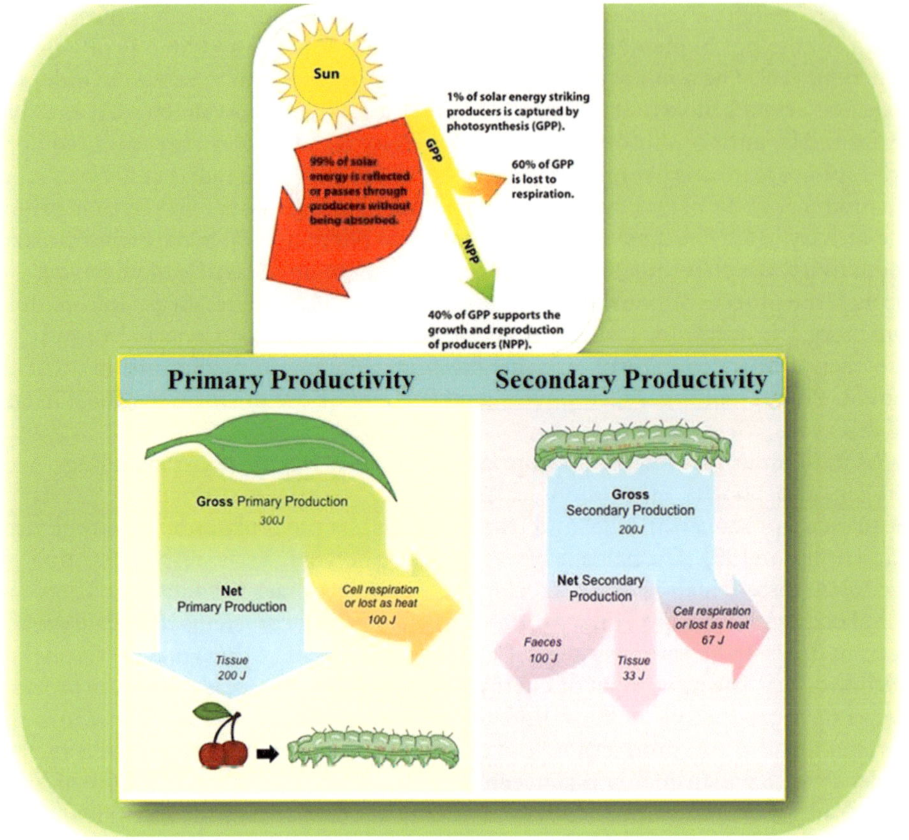


**Figure 4.**
*Schematic representation of productivity of aquatic system.*

$$Net\ Primary\ Productvity = \frac{LB - IB}{\mathrm{T}} \mathrm{X} \frac{0.375}{PQ} \mathrm{X}\, 1000 mg\,/\,\mathrm{L}\,/\,\mathrm{h}$$

$$Respiration = \frac{IB - DB}{\mathrm{T}} \mathrm{X} \frac{0.375}{PQ} \mathrm{X}\, 1000 mg\,/\,\mathrm{L}\,/\,\mathrm{h}$$

Where:
LB = Light bottle,
DB = Dark bottle,
IB = Initial bottle,
T = Time of incubation,
PQ = Photosynthesis Quotient = 1.25,
RQ = Respiratory Quotient =1 and.
The value 0.375 represents a constant to convert Oxygen value to Carbon Value.

## 9. Conclusion

The zooplankton's role in lentic and lotic water system to improve the water quality, act as basal organism in food chain and maintain the environmental conditions.

The existence and richness of zooplanktons relationship associated with water quality in both the water system. Zooplankton communities such as cladocerans, copepods, rotifers, etc., are held responsible to set biological parameters due to their central role in aquatic habitat in food webs along with the phytoplankton's (primary producers) and reach to higher level trophic system and to become important in energy flow and in nutrient cycle of aquatic systems. The many research study showed the correlation between dominant species along with the measured environmental parameters, whereas many tools, methods and protocols like cluster analysis, population metrics, limnology, etc., will help us to know the lentic and lotic water assemblages of zooplankton. This chapter emphasised that zooplankton community establish the structure of water system, environmental conditions and reflects the health of lentic and lotic ecosystem, hence they play significant role in biomonitoring, biodegrading and bioremedial of any water ecosystem health.

## Author details

N. Eramma[1], H.M. Lalita[2], S. Satishgouda[3], S. Renuka Jyothi[4], C.N. Venkatesh[5] and Sharangouda J. Patil[6]*

1 Department of Microbiology, Acharya Bangalore B-School, Bengaluru, Karnataka, India

2 Department of Zoology, University College of Science, Tumkur University, Tumkur, Karnataka, India

3 Department of Zoology, Government Science College, Chitradurga, Karnataka, India

4 Department of Life Sciences, Jain (Deemed to be) University, Bengaluru, Karnataka, India

5 Department of Zoology, S.J.M. Arts, Science and Commerce Degree College, Chitradurga, Karnataka, India

6 Department of Zoology, NMKRV College for Women (Autonomous), Bengaluru, Karnataka, India

*Address all correspondence to: shajapatil@gmail.com

## References


[1] Ahmed MH, Naida NB, El-Hamed A, Shalby NI. Impact of Physico-chemical parameters on composition and diversity of zooplankton Community in Nozha Hydrodrome, Alexandria, Egypt. Egyptian Journal of Aquatic Biology and Fisheries. 2017;**21**(1):49-62

[2] Abo-Taleb H. Importance of plankton to fish community. In: Biological Research in Aquatic Science. London, UK: IntechOpen; 2019. 10.5772/intechopen.85769

[3] Moss B. Ecology of Fresh Waters. Man and Medium-Past to Future. Oxford: Blackwell Science; 1998

[4] Arribas P, Velasco J, Abellan P, Sanchez-Fernandez D, Andujar C, Calosi P, et al. Dispersal ability rather than ecological tolerance drives differences in range size between lentic and lotic water beetles (Coleoptera: Hydrophilidae). Journal of Biogeography. 2012;**39**:984-994. DOI: 10.1111/j.1365-2699.2011.02641.x

[5] Hutchinson GE. A Treatise on Limnology. New York: Wiley; 1957

[6] Damm S, Dijkstra KDB, Hadrys H. Red drifters and dark residents: The phylogeny and ecology of a Plio-Pleistocene dragonfly radiation reflects Africa's changing environment (Odonata, Libellulidae, Trithemis). Molecular Phylogenetics and Evolution. 2010;**54**:870-882. DOI: 10.1016/j.ympev.2009.12.006

[7] Hof C, Brandle M, Brandl R. Lentic Odonates have larger and more northern ranges than lotic species. Journal of Biogeography. 2006;**33**:63-70. DOI: 10.1111/j.1365-2699.2005.01358.x

[8] Reddy MT, Sivaraj N, Kamala V, Pandravada SR, Sunil N, Dikshit N. Classification, characterization and comparison of aquatic ecosystems in the landscape of Adilabad District, Telangana, Deccan region, India. Open Access Library Journal. 2018;**5**:e4459. DOI: 10.4236/oalib.1104459

[9] Pal S, Chakraborty K. Quantification of zooplanktons from two selected freshwater bodies of Cooch Behar District of West Bengal, India. Elixir Applied Zoology. 2016;**101**:44085-44088

[10] Ziauddin G. Study of Limnological status of two selected floodplain wetlands of West Bengal. Research Square. 2021:1-68. DOI: 10.21203/rs.3.rs-443104/v1

[11] Joseph B, Yamakanamardi SM. Monthly changes in the abundance and biomass of zooplankton and water quality parameters in Kukkarahalli Lake of Mysore, India. Journal of Environmental Biology. 2010;**32**:551-557

[12] Sharma P, Giri A. Productivity evaluation of lotic and lentic water body in Himachal Pradesh, India. MOJ Ecology & Environmental Sciences. 2018;**3**(5):311-317

[13] Meena K, Dube P. A critical review of zooplankton studies of lentic water bodies of India. International Journal of Environmental Sciences. 2018;**7**(3):79-83

[14] Ruhl N, Haban D, Czajkoski C, Grove M, Richmond CE. Community composition of zooplankton exported from a shallow Polymictic reservoir linked to wind conditions. Peer J. 2019;**7**:e7611. DOI: 10.7717/peerj.7611

[15] Kalff J. Limnology: Inland Water Ecosystems. New Jersey: Prentice Hall; 2002

[16] Wetzel RG. Limnology: Lake and River Ecosystems. Third ed. New York: Academic Press; 2001

[17] Smith RL. Ecology and Field Biology. Natural History of Nova Scotia. 3rd ed. New York: Harper & Row; 1980. Vol. I. 1990. pp. 428-445

[18] Tokeshi M, Pinder LCV. Microhabitats of stream invertebrates on two submersed Macrophytes with contrasting leaf morphology. Ecography. 1985;**8**(4):313-319. DOI: 10.1111/j.1600-0587.1985.tb01184.x

[19] Brüchner-Hüttemann H, Ptatscheck C, Traunspurger W. Meiofauna in stream habitats: Temporal dynamics of abundance, biomass and secondary production in different substrate microhabitats in a first-order stream. Aquatic Ecology. 2020;**54**:1079-1095. DOI: 10.1007/s10452-020-09795-5

[20] Padmanabha B. Comparative study on the hydrographical status in the lentic and lotic ecosystems. Global J Ecol. 2017;**2**(1):015-018

[21] Wan Maznah WO, Intan S, Sharifah R, Lim CC. Lentic and lotic assemblages of zooplankton in a tropical reservoir, and their association with water quality conditions. International journal of Environmental Science and Technology. 2017;**15**(3):533-542

[22] Bhardwaj S, Mishra H. Aquatic Ecosystem: Freshwater Ecosystem: Part 2. New Delhi: MHRD. Govt. of India; 2020. pp. 1-33

[23] Cushing CE, Allan JD. Streams: Their Ecology and Life. San Diego: Academic Press; 2001. p. 366

[24] Giller S, Malmqvist B. The Biology of Streams and Rivers. Oxford: Oxford University Press; 1998. p. 296

[25] Hafiz BZ, Hafiz IZ. Review on Nationwide surveillance for phytoplankton ecology in India. International Journal of Current Research and Academic Review. 2018;**6**(10):1-10

[26] Clarke A, Peck LS. The physiology of polar marine Zooplankton. In: Sakshaug E, Hopkins CCE, Britsland NA, editors. Proceedings of the pro Mare Symposium on Polar Marine Ecology. Vol. 10, Issue 2. Trondheim: Polar Research; 1991. pp. 355-369

[27] Saidin ANB. Water Quality and Zooplankton Community Structure of Tembat River, Hulu Terengganu [thesis]. School of Biological Sciences; Universiti Sains Malaysia. 2016

[28] Hötzel G, Croome R. A Phytoplankton Methods Manual for Australian Freshwaters. LWRRDC Occasional Paper 22/99, Land and Water Resources Research and Development Corporation, Canberra. 1999

[29] Goldman CR, Home AJ. Limnology. London: McGraw-Hill Intl. Book Co.; 1983. p. 464

[30] Alcaraz M, Albert C. Large Zooplankton: Its Role in Pelagic Food Webs. In Book: Fisheries and Aquaculture – Vol. V; EOLSS: UNESCO; 2007. pp. 01-08

[31] Stockwell JD, Doubek JP, Adrian R, et al. Storm impacts on phytoplankton community dynamics in lakes. Global Change Biology. 2020;**26**(5):2756-2784. DOI: 10.1111/gcb.15033

[32] Bera B. Using zooplankton community to assess water quality and trophic condition of Lake Saheb bandh, Purulia, West Bengal, India. International Journal of Aquatic Science. 2021;**12**(02):4471-4498

[33] Delince G. The Ecology of the Fish Pond Ecosystem with Special Reference

to Africa. Netherland: Pergamon Press; 1992. pp. 220-230

[34] Odum EP. Fundamentals of Ecology. 3rd ed. Philadelphia: W.B. Saunders Company; 1971. pp. 1-574

[35] Goldsborough LG, Robinson GGC. Pattern in wetlands. In: Stevenson RJ, Bothwell ML, Lowe RL, editors. Algal Ecology: Freshwater Benthic Ecosystems. Academic Press: San Diego; 1996. pp. 77-117

[36] Verma BS, Srivastava SK. Study of factors affecting phytoplankton primary productivity in a pond of Patna, Bihar, India. Nature Environment and Pollution Technology An International Quarterly Scientific Journal. 2015;**15**:291-296

[37] Kumar A. Studies on monthly and seasonal variations in primary productivity of glacial fed mountainous Goriganga River in Kumaun Himalaya, Uttarakhand, India. International Research Journal of Biological Sciences. 2015;**4**:53-56

[38] Garder TG, Green HH. Investigation of the Production of Plankton in the Osio Ford. Rapp. Process- Reunions Council Pernt. International Exploration. 1927;**42**:1-48

Chapter 2

# Perspective Chapter: *Daphnia magna* as a Potential Indicator of Reservoir Water Quality – Current Status and Perspectives Focused in Ecotoxicological Classes Regarding the Risk Prediction

*Sara Rodrigues, Ivo Pinto, Sandra Nogueira and Sara C. Antunes*

## Abstract

Several types of stressors come into natural water bodies, degrading their quality, and having harmful effects on aquatic biota. As a result, many attempts have been made to develop complementary techniques to those imposed by the Water Framework Directive (WFD) to improve the water quality assessment strategy in a shorter time and be more faithful to the components and contaminants of the ecosystem. *Daphnia magna* has been extensively used as a model organism for ecotoxicity testing, and its ecotoxicological responses to several disturbance factors have been being well characterized. From this perspective, this work aimed to evaluate the applicability of the feeding bioassays with *D. magna*, as well as early distress tools (biochemical biomarkers), in the assessment of water quality of natural waters of reservoirs. Samplings were performed in several sites in three Portuguese reservoirs and were conducted in the spring of 2020. Bioassays and biomarkers results showed sensitivity to different reservoirs since the sites are minimally or moderately impacted. Biological responses can be related to several environmental factors, such as surrounding areas, seston composition, and chemical analysis (WFD), among others not quantified. This set of biological responses has presented good concordance with the ecological potential of the reservoirs.

**Keywords:** water framework directive, lentic ecosystems, model species, bioassays and biomarkers, ecotoxicity classes

## 1. Introduction

Over the years, increasing demographic pressures have contributed to an exponential increase in contaminants in the water ecosystems, associated with the intensification of agricultural and industrial activities. Some of these contaminants have raised particular concern and have been classified as specific pollutants and priority substances, identified in the Directive 2013/39/EU of the European Parliament [1] and the Agência Portuguesa do Ambiente (APA) [2]. To perform the ecological assessment of aquatic ecosystems, there are two major comprehensive frameworks worldwide, using multiple lines of evidence (LoE), with special emphasis on ecological data (biological communities): the Water Framework Directive (WFD), adopted in Europe through the Directive 2000/06/EC, and the Ecological Risk Assessment (ERA), adopted for example by the US Environmental Protection Agency. WFD and ERA approaches integrate information from different LoE, as the extent of the application of ecotoxicological evidence clarifying cause-effect relationships; the availability of expert judgment in the fine-tuning of sampling practices, strategic analysis, data interpretation, and decision procedures; on the practical meaning of the concept of LoE integration ("one-out, all-out" principle versus integrated risk quantification) [3]. The WFD is an extensive legislative framework for the protection of ground and surface waters in Europe, with defined Environmental Quality standards (EQS) for several parameters [biological, chemical, physicochemical (supporting biological elements), and hydromorphological elements] that must be complied by the different member states, and also advise additional monitoring of substances of national or regional interest [1]. However, it is estimated that a very high number of substances are present in the environment, and for Europe, it is estimated the existence of more than 100,000 compounds [4, 5]. Therefore, the definition of these lists of compounds with environmental concern, although relevant, becomes unreliable, in terms of representation in aquatic ecosystems, since complex mixtures of chemicals occur (e.g., emerging pollutants, metabolites, and transformation products). A list of priority substances, which represent a significant risk to or via the aquatic environment at the EU level, will have to be reexamined by the European Commission (EC) and should not exceed 4 years. In recent years, there has been a growing agreement among authorities and scientists that the tools currently used and proposed by the WFD for water quality assessment require a review to achieve a clearer and future-proof methodology [6, 7]. In this context, ERA is reflected as a complementary alternative for the bioassessment of the quality of freshwater, such as reservoirs. This approach considers some valuable WFD principles and metrics but, at the same time, includes complementary methods, one of which is the incorporation of effects-based tools (e.g., ecotoxicological assays; biomarkers in organisms) for a better assessment of cause-effect relationships; reflecting an effective integration of distinct LoEs (e.g., chemical, ecological and ecotoxicological) [3].

In environmental terms, pollution occurs due to a complex mixture of organic and inorganic compounds that can result in lethal and sub-lethal effects on aquatic organisms, associated with potentially significant losses of habitat and biodiversity. Currently, the assessment of water quality, using organisms as bioindicators of water quality has been widely used, since the biological responses integrate the complex influence of the stressing agents [4, 6, 8–10], in addition to the complex mixture of compounds that occur in the ecosystem under analysis. According to several studies, the effect-based water quality assessment (e.g. biological responses of organisms to natural waters) has been successful in the identification of ecotoxicological risks in surface waters and the ranking of locations based on these risks, namely for natural water bodies (e.g. rivers, transitional or coastal waters) [5, 6, 11], but also to heavily modified water bodies (e.g. reservoirs) [8–10]. Bioassays with *Daphnia* sp. (mostly *Daphnia magna*) are regularly used in ecotoxicological studies because they have high fertility values, easy to maintain in laboratory conditions, ubiquitous, and important bioindicators for aquatic environments due to their sensitivity to contaminants and position (trophic level: primary consumers) in the aquatic food webs [6, 9, 11].

To answer the research needs outlined above, the present study aimed to apply effect-based approaches (individual and biochemical responses of *D. magna* to natural waters) in the water quality assessment of Portuguese reservoirs, defining classes of disturbances and ranges of ecotoxicological potential. This work has been divided into two parts with specific objectives. Part 1 of this study was presented in [9], which demonstrated that biochemical parameters (metabolism, oxidative stress, and lipid peroxidation biomarkers) improved the sensitivity of the biomonitoring strategy using bioassays with the standard species *D. magna*, in the assessment of the ecological quality of water reservoirs, in different sampling periods (Autumn 2018 and Spring, Autumn 2019). The biochemical parameters revealed sensitivity in the evaluation of effects incited by exposure to natural waters from reservoirs, making them useful and reliable in this type of evaluation. According to the results of [9], the biomarker indicative of lipid peroxidation (levels of thiobarbituric acid reactive substances—TBARS) on *D. magna* represented a consistent tool for evaluation of water quality. This result reflected, in part, the prooxidative state of organisms, food performance, and possible stress scenarios, mainly due to the components of seston (e.g., quantity and quality of phytoplankton) and chemical contamination.

Part 2, the focus of the here-presented study, is intended to evaluate whether effect-based methods can be applied in natural waters quality assessment of reservoirs, by the definition of disturbance classes and ecotoxicological potential values. The integration of physical and chemical and effects-based monitoring approaches can complement and improve the water quality assessment strategies in the future, with the main objective of a nontoxic environment. For this, several ecotoxicological tools were applied, to gain insight into the ecotoxicological potential of the reservoirs understudy and compare the ecological potential with tools proposed by the WFD, as well as with previous studies, in the same areas under study. To this end, the present approach combined the evaluation of the individual (feeding rates) and cellular/molecular responses in *D. magna* after acute exposure to natural waters [e.g., biochemical biomarkers of oxidative response as activities of catalase (CAT) and glutathione S-transferases (GSTs), the latter also involved in the biotransformation process, lipid peroxidation (LPO measured as TBARS levels), and acetylcholinesterase (AChE) activity, involved in neurotransmission process].

## 2. Material and methods

### 2.1 Study areas, water sampling, and physicochemical parameters quantifications

Three reservoirs were selected for conducting this study (**Figure 1**): Miranda (M) and Pocinho (P) are main course reservoirs and belong to the hydrographic basin of the Douro river; and Aguieira (Ag) which is a northern reservoir that belongs to the Mondego hydrographic basin.

The purpose for defining these reservoirs and respective sampling sites was based on previous studies of our research group [8–10, 12–14], where the water quality was assessed, in the last years, using different indicators and methodologies complementary to the WFD.

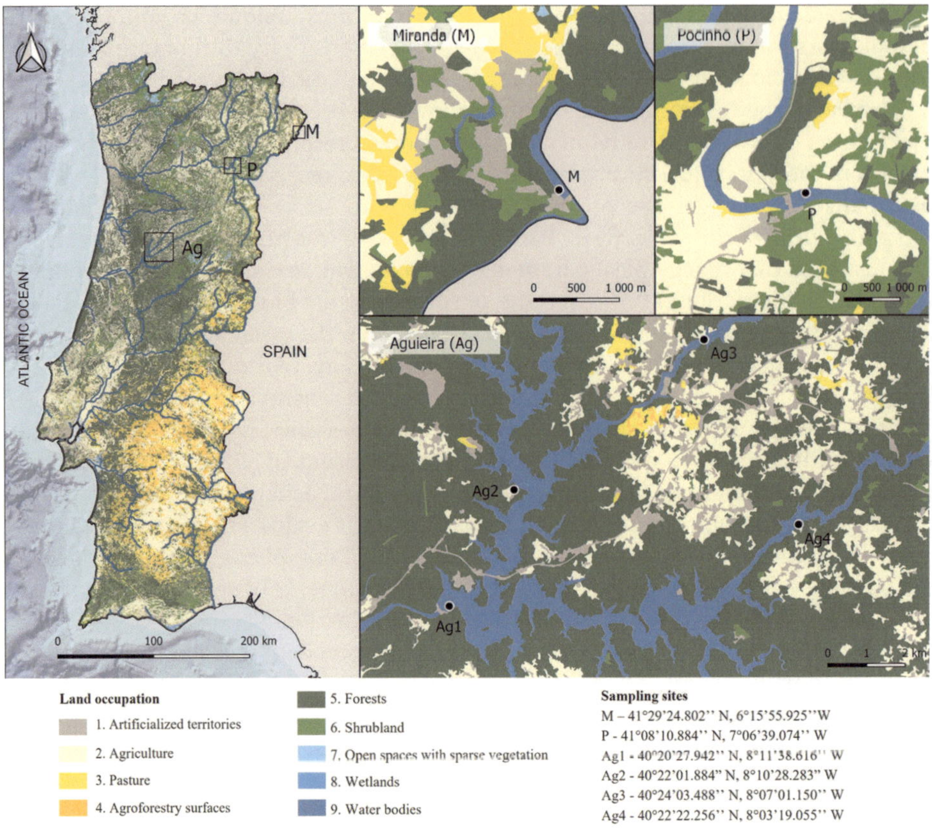


**Figure 1.**
*Map of sampling areas (Miranda, Pocinho and Aguieira reservoirs) with the location of the sampling sites. The different colors represent the 1st level of detail of land occupation according to the land use report (2018).*

Water samples were collected during the spring of 2020, in six sites of the three reservoirs (**Figure 1**): one site in Miranda, one site in Pocinho, and four sites in Aguieira. The sites were well-defined based on previous works conducted in these reservoirs [8–10, 12–14].

*In situ*, the abiotic parameters pH, dissolved oxygen (mg/L and %), conductivity (μS/cm), and temperature (°C) were measured with a multiparameter probe (Multi 3630 IDS SET F). For conducting chemical analysis [e.g., nutrients, specific pollutants, and priority substances] and bioassays, 5 L of water were collected at each sampling site and transported to the laboratory at 4°C and in the dark. Chemical analyses were carried out within a maximum period of 48 hours after collection. *D. magna* assays were started within a maximum period of 24 hours, after sample collection.

## 2.2 Chemical analyses

A set of specific pollutants and priority substances were measured, according to the recommendations defined in the Agência Portuguesa do Ambiente [2] and the Directive 2013/39/EU [1]. Nitrites ($NO_2^-$) and nitrates ($NO_3^-$) were quantified by liquid chromatography of ions, as dissolved anions [15]. Total Kjeldahl nitrogen ($N_{Kj}$) determination was performed by the Kjeldahl nitrogen method after mineralization with selenium [16]. Calcium and magnesium determinations were effectuated by ion chromatography, as dissolved cations [17]. For the elements, total phosphorus ($P_{total}$), arsenic, cadmium, copper, mercury, nickel, lead, and zinc, the analysis was performed by the application of inductively coupled plasma mass spectrometry (ICP-MS) [18]. Pesticides and polycyclic aromatic hydrocarbons were not quantified in our study, because, according to previous studies [8–10, 12–14] the values of these specific pollutants and priority substances (quantified in autumn of 2018) were below the detection limits of the analytical method, in addition to, that no significant changes in the areas adjacent to the reservoirs were documented during the last years (2019, 2020).

### 2.3 Biological parameters by WFD—Ecological quality ratio (EQR) for phytoplankton

The phytoplankton community characterization was performed according to the Instituto da água I.P. [19] and Agência Portuguesa do Ambiente (APA) [2] guidelines and briefly described in [9]. For the determination of the ecological potential (EP), the results were expressed in an EQR, determined according to the WFD approach. According to APA [2], the EQS used in the classification of the biological quality (EQR) for Miranda and Pocinho reservoirs was carried out based on the typology "main course". For main course typology, taking into account the biological elements proposed in the WFD to Portuguese reservoirs, the EP is only classified into two classes: moderate or less, and good or more (**Table 1**). Aguieira is a northern type of reservoir, and the EP is classified into four classes: Good or more, Moderate, Poor, or Bad (**Table 1**).

### 2.4 Water treatments

The water collected in each sampling site of each reservoir was processed in 3 treatments, namely: NF (Non-Filtered water with all components present in the sample); F1 (water filtered through a Whatman GF/C filter with 1.2 μm porosity); and F2 (water filtered through a sterile filter system with a porosity of 0.22 μm) as already defined in previous studies of our group [10, 12].

### 2.5 Test organisms

#### *2.5.1 Culture maintenance of* D. magna

Successive generations of monoclonal cultures of *D. magna* were continuously kept in controlled laboratory conditions of $16^h$ light and $8^h$ dark photoperiod and temperature of 20 ± 2°C. Cultures were renewed every 2 days and were maintained in synthetic water medium "ASTM hard water" [20], supplemented with a standard organic additive, *Ascophyllum nodosum* extract [21], to provide essential microelements to *Daphnia*. *D. magna* were fed with *Raphidocelis subcapitata* at a rate of $3.0x10^5$ cells/mL/day. For conducting the bioassays, daphnids with 4 or 5 days, born between the 3rd and 5th broods were used.

#### *2.5.2* D. magna *feeding rate assays*

*D. magna* feeding rate assays were conducted according to [22] with some adaptations described in [9]. For each water sample, bioassays were performed on 6-well

| | | | Main course | | North | | | |
|---|---|---|---|---|---|---|---|---|
| | | **EQS** | **M** | **P** | **Ag1** | **Ag2** | **Ag3** | **Ag4** |
| **Physical and chemical elements** | Temp (°C) | — | 19.0 | 22.8 | 21.9 | 20.5 | 20.7 | 22.4 |
| | Cond (μS/cm) | — | 438 | 268 | 74 | 85 | 90 | 73 |
| | pH | 6–9 [2] | 8.6 | **9.2** | **9.6** | **9.7** | 9.0 | **9.4** |
| | $O_2$ (mg/L) | ≥5 [2] | 11.0 | 15.9 | 12.9 | 14.2 | 12.4 | 13.3 |
| | $O_2$ (%) | 60–120 [2] | 124.1 | 185.0 | 150.2 | 160.1 | 141.0 | 156.5 |
| | $NO_2^-$ (mg $NO_2^-$/L) | | 0.16 | 0.05 | 0.04 | 0.04 | 0.05 | 0.01 |
| | $NO_3^-$ (mg $NO_3^-$/L) | ≤ 25 [2] | 7.4 | 3.5 | 2.7 | 2.2 | 3.3 | 0.6 |
| | $N_{Kj}$ (mg N/L) | — | <0.5 | 0.7 | <0.5 | 0.7 | <0.5 | <0.5 |
| | $P_{total}$ (mg P/L) | ≤ 0.05 [2] | **0.13** | **0.09** | 0.02 | 0.03 | **0.08** | 0.03 |
| **Specific Pollutants** | As (μg/L) | 50 [2] | 2.13 | 3.30 | 1.41 | 1.31 | 1.62 | 2.83 |
| | Zn (μg/L) | 7.8 [2] | **26.6** | **80.7** | **32.2** | **27.7** | **28.9** | **24.7** |
| | Cu (μg/L) | 7.8 [2] | 1.84 | 1.59 | 1.86 | 2.09 | 2.92 | 1.70 |
| **Priority substances** | Cd (μg/L) | 0.45 [1] | 0.01 | 0.02 | 0.02 | 0.02 | 0.02 | 0.01 |
| | Hg (μg/L) | 0.07 [1] | **1.02** | **1.06** | **1.85** | **1.57** | **1.33** | **0.63** |
| | Ni (μg/L) | 34 [1] | 2.1 | 2.4 | 2.3 | 1.7 | 1.6 | 1.5 |
| | Pb (μg/L) | 14 [1] | 0.5 | 0.5 | 0.4 | 0.3 | 0.4 | 0.3 |
| | Ca (mg/L) | — | 63 | 40 | 3 | 4 | 4 | 3 |
| | Mg (mg/L) | — | 10 | 6 | 1 | 1 | 2 | 1 |
| **Ecological potential (chemical and physico-chemical elements)** | | | Moderate | Moderate | Moderate | Moderate | Moderate | Moderate |
| **Biological (Phytoplankton EQR)** | **North** [2]<br>[1.0–0.60] – Good or more<br>[0.6–0.4] – Moderate<br>[0.4–0.2] – Poor<br>[0.2–0] – Bad | | 0.16 | 0.12 | 0.77 | 0.37 | 0.33 | 0.61 |

| | | Main course | | North | | | |
|---|---|---|---|---|---|---|---|
| | EQS | M | P | Ag1 | Ag2 | Ag3 | Ag4 |
| | **Main course** [2]<br>≥0.17 – Good or more<br>< 0.17 – Moderate or less | | | | | | |
| **Ecological potential (biological)** | | Moderate or less | Moderate or less | Good or more | Poor | Poor | Good or more |

*Sampling sites: Miranda – M, Pocinho – P and Aguieira – Ag1 to Ag4. Temperature (Temp), Conductivity (Cond), pH, Dissolved oxygen ($O_2$), nitrites ($NO_2^-$), nitrates ($NO_3^-$), Ammonium ($NH_4^+$), Total Kjeldahl nitrogen ($N_{Kj}$), Total phosphorus ($P_{total}$), Arsenic (As), Zinc (Zn), Copper (Cu), Cadmium (Cd), Mercury (Hg), Nickel (Ni), Lead (Pb), Calcium (Ca), Magnesium (Mg), Iron (Fe), Manganese (Mn), and Ecological Quality Ratio (EQR).*

**Table 1.**
*Results of the physical and chemical parameters, and specific pollutants and priority substances concentrations of Portuguese reservoirs. The bold values represent the values outside of the environmental quality standards (EQS). The biological parameter phytoplankton and respective EQS for the main course and north reservoirs are also presented.*

plates, where each plate corresponded to specific water treatment (NF, F1, or F2). For each water treatment and control (ASTM hard water medium), 5 replicate wells with 5 *D. magna* individuals, and a blank well (water sample with *Raphidocelis subcapitata* without daphnids) were performed. The blank well is performed to account for the potential algal growth during the assay period. Mortality was also considered in this study. Feeding rate results were expressed according to [23]. The percent inhibition in feeding rate (% $I_{FR}$), relatively to the control, was calculated for each water treatment (NF, F1, and F2) as follows:

$$\%I_{FR} = [(FRc - FRt)/FRc] \times 100 \quad (1)$$

where:
% $I_{FR}$: percent inhibition of feeding rate;
FRc: mean value for feeding rate in the control group;
FRt: value for feeding rate for the water treatment.

At the end of the feeding rate assays, pools of organisms from each treatment were preserved for posterior biochemical determinations (oxidative stress, lipid peroxidation, and neurotransmission biomarkers) and stored in microtubes at—80°C until analyses were performed.

*2.5.3 Biochemical determinations*

For determination of biomarkers of oxidative stress [catalase (CAT) and isoenzymes glutathione S-transferases (GSTs) activities] and levels of lipid peroxidation (LPO) [levels of thiobarbituric acid reactive substances (TBARS)], samples were thawed on ice, and a 1 mL of ice-cold phosphate buffer (50 mM, pH = 7.0 with 0.1% of Triton X-100) was added to each biological sample. Samples were sonicated for 20 s and centrifuged at 14,000 rpm, for 10 min, at 4°C, in a refrigerated centrifuge (Eppendorf 5810R). The supernatant fraction was divided into aliquots and used to perform the biochemical analyses. For the quantification of acetylcholinesterase (AChE) activity, the samples were homogenized with a sonicator, in a volume of 750 μL of ice-cold phosphate buffer (0.1 M phosphate buffer, pH = 7.2), and centrifuged at 6000 rpm for 3 min. The supernatants after centrifugation were collected and used for AChE activity determinations.

All biochemical analyses were adapted to 96-well microplates [9, 24], and spectrophotometric readings were performed in a microplate reader Thermo Scientific, model Multiskan GO (version 1.00.40), with Skanlt Software 3.2.

The total soluble protein concentration of samples was performed according to the Bradford method [25], using a standard of γ-globulin 1 mg/mL. This method is based on the binding of a dye (Bradford's reagent) to the total soluble proteins, forming a stable complex [24]. This determination permits expressing the enzymatic activities and TBARS levels, as a function of the total soluble protein content of the samples.

CAT is an antioxidant enzyme with peroxidic activity and is responsible for the decomposition of hydrogen peroxide ($H_2O_2$) in $H_2O$ + $O_2$, where $H_2O_2$ consumption occurs with the oxidation of hydrogen donors (phenols, formic acid, and methanol) [24]. The method consists of the monitoring of this decomposition and was performed according to [26]. The enzymatic activity was expressed in nmoles of $H_2O_2$ consumed, per minute, per milligram of total soluble protein.

The GSTs activity was quantified according to [27]. GSTs catalyze the conjugation of glutathione in its reduced form (GSH) with the substrate 1-chloro-2,4-

dinitrobenzene (CDNB), forming a thioether, whose formation was observed by measuring the increase of the absorbance [24]. Enzyme activities were expressed in mmol of thioether produced, per minute, per milligram of total soluble protein.

LPO was measured through the determination of the levels of TBARS, according to [28], which measures the absorbance of the complex resulting from products of oxidative free radical attack to membrane lipids, with thiobarbituric acid. Results were expressed as millimoles of malondialdehyde (MDA) and MDA-like compounds equivalents, per mg of total soluble protein.

The quantification of the AChE activity was performed by the Ellman method [29]. This enzyme is responsible for the degradation of the synthetic substrate acetylthiocholine into acetate + thiocoline [24], which occurs with the increase in the yellow color produced when thiocoline is complexed with dithiobis nitrobenzoate (DTNB). The enzymatic activity was expressed as nmol of the complex formed, per minute, per milligrams of total soluble protein.

### 2.6 Water ecotoxicological assessment

At the end of all quantifications of enzyme activities and LPO levels, the percent inhibition in each biochemical determination (% $I_X$), comparatively to the control, was calculated for each water treatment (NF, F1, and F2) as follows:

$$\%I_X = [(Xc - Xt)/Xc] \times 100 \tag{2}$$

where:

% $I_X$: percent inhibition of: CAT activity (% $I_{CAT}$), GSTs activity (% $I_{GSTs}$), TBARS levels (% $I_{TBARS}$) or AChE activity (% $I_{AchE}$);

Xc: mean value for CAT activity, GSTs activity, TBARS levels, or AChE activity in the control group;

Xt: value for CAT activity, GSTs activity, TBARS levels, or AChE activity for the water treatment.

### 2.7 Statistical analyses

The data from all test variables (the percent inhibition of feeding rate, CAT and GSTs activities, TBARS levels, and AChE activity) were previously analyzed to assure normality and uniformity of variance (Shapiro–Wilk and Levene tests, respectively). All parameters were analyzed by analysis of variance (one-way ANOVA), followed, when significant differences were detected ($p < 0.05$), by a Tukey test to discriminate differences between treatments (NF, F1, and F2). The data are presented as mean and respective standard errors. The analyses were performed using software SPSS Statistics (version 26) and Sigmaplot (version 11.0).

## 3. Results and discussion

### 3.1 General physicochemical characteristics and trace elements concentrations (chemical analysis) in the water samples

**Table 1** summarizes the results of physicochemical parameters and chemical analyses including the quantifications of the concentrations of the specific pollutants and priority substances, measured for each site over the sampling period, as recommended by APA [2] and European Parliament and the Council [1]. According to the physicochemical parameters used in the WFD, for Portuguese heavily modified and artificial water bodies, only the pH, $O_2$, $NO_{3^-}$ and $P_{total}$ have an environmental quality standard (EQS) values established for a good ecological potential (GEP).

In general, water samples from the three reservoirs were characterized by a basic pH (values range from 8.6 and 9.7, **Table 1**), with almost values above the EQS (> 9.0, except M and Ag3). The water pH is an important parameter as it can determine the solubility and biological availability of nutrients, but also metals [30]. Dissolved oxygen (%) showed values above the maximum of 120%, in all sites studied. Considering the electrical conductivity, the values ranged between 73 and 438 μS/cm. Higher values of this parameter were registered in the reservoirs belonging to the Douro river basin, Miranda and Pocinho (> 260 μS/cm). In different Aguieira sites, the values showed a low variation between the locations of 73–90 μS/cm. The sites M, P, and Ag3 showed higher contents of nutrients (mainly total phosphorus) when compared with the values of EQS (**Table 1**). Two types of reservoirs are referred to in the work of [31]: Type 1—lowland "run-of-river" reservoirs located in the main rivers (e.g., Douro), at lower altitudes, had larger catchments, lower residence time, and were higher in mineral content (hardness and conductivity), than Type 2, which are deeper high altitude reservoirs (e.g., Mondego). Considering this distinction, Miranda and Pocinho are reservoirs of Type 1 and Aguieira is Type 2 (for more information see [31]), and in fact, our physicochemical results are supported by these assumptions (**Table 1**). Higher nutrient concentrations ($P_{total}$) were observed at Miranda and Pocinho sites than at the Aguieira sites (**Table 1**). According to [31], Type 1 reservoirs are more nutrient-rich (total phosphate and nitrates due to more extensive agriculture and intensive) than Type 2, corroborating our results. If we consider land occupation (**Figure 1**) in the area surrounding the sampling site of Miranda, the water pressures are associated with the artificialized territories and forests. For Pocinho, agriculture is highly representative in terms of land occupation. The land occupation in the Mondego river basin area, where the Aguieira reservoir is located, has the surrounding areas mainly represented by forests, agricultural areas, and artificialized territories (**Figure 1**). Kroll et al. [32] show a solid association between land occupation (urban, agriculture, and forest areas) and nutrients indices in nearby aquatic ecosystems. These findings demystify and support some of the results (e.g., nutrient levels) presented here, as well as work previously developed in these same locations [9].

Concerning the metals, only mercury (Hg) and zinc (Zn) exceeded the EQS (**Table 1**) defined by the Directive 2013/39/EU of the European Parliament [1] and APA [2], respectively. Hg was present in concentrations above 0.07 μg/L (EQS) at all sites (> 0.63 μg/L). For Zn, concentrations above 7.8 μg/L (EQS) were quantified in all locations of the three reservoirs (> 24.7 μg/L). Several metals such as mercury and zinc (among others) can be highly toxic even in residual quantities [33]. Hg is an important pollutant of water throughout the world, and several human activities are linked to Hg pollution (silver and gold mining, coal combustion, and dental amalgams), and is known to be an inhibitor of enzymes' activities [9, 30, 34, 35]. The speciation of Zn in water is modulated by pH and dissolved organic matter, which normally binds most of the aqueous zinc [30]. Zn concentrations in natural waters span six orders of magnitude and are strongly influenced by human activities [30]. There are a comprehensive set of proteins that function as transporters, chelators, and molecular sensors for Zn, and are involved in the regulation of Zn uptake by homeostatic processes that are partially understood. However, several studies have proposed theories to explain how zinc compounds affect aquatic animals [30]. However, inter- and intra-specific differences cannot be disregarded, as well as doses and exposure times.

Anthropogenic activities have been found to contribute more to environmental contamination (e.g., water eutrophication which was recognized in the middle and late stages of the twentieth century) due to the everyday manufacturing of materials to meet the demands of the population [36, 37], in its various aspects that include, agriculture, industry, and urban areas. As mentioned by [37] human interference is to a greater extent caused by social and economic pressures, which are associated with the largest changes that occurred in agricultural and forest areas as a result of the extensification of agriculture, deforestation, afforestation, and urbanization. In Europe, these are the trends observed over the last years [38, 39].

### 3.2 Ecological quality ratio (EQR) for phytoplankton

In general, the phytoplankton EQR (**Table 1**) shows that the Miranda and Pocinho reservoirs had the worst water quality (moderate or lower), taking into account the defined classes for the main water course typology. The Aguieira reservoir tended to have low water quality, with Ag2 and Ag3 being the most problematic sites with the lowest EQR values recorded. All reservoirs were characterized as eutrophic, especially due to high concentrations of $P_{total}$ recorded over the last few years [8–10, 12–14, 31, 35, 40], a condition also observed in the present study (**Table 1**). The bioavailability of nutrients such as phosphorus favors the overgrowth of phytoplanktonic communities [8–10, 12–14], namely, cyanobacteria organisms, that were already associated with poor water quality and recurrently reported blooms was been in these reservoirs [8–10, 12–14]. The most prevalent and main group of cyanobacteria detected in all reservoirs was *Microcystis*. However, *Anabaena*, *Woronichinia*, and *Pseudanabaena* (and others) were also detected, but in a much smaller percentage. The percentage of cyanobacteria detected in Miranda and Pocinho was 54.05, and 57.32%, respectively. Lower percentages of cyanobacteria were observed in the Aguieira sites, namely, 1.07% (Ag1), 4.75% (Ag2), 10.75% (Ag3), and 27.30% (Ag4). However, other phytoplankton indicators are included in the assessment of this reservoir typology (e.g., Algae Group Index (AGI) [8, 10], which is strongly interfering with the final classification of the Ag4 site.

### 3.3 Water ecotoxicological assessment

The proposal of classes of disturbances (defined by colors) and ecotoxicity results for *D. magna*, after exposure to treatments of the natural waters of Miranda, Pocinho, and Aguieira reservoirs, are presented in **Figures 2–4**, respectively. For each figure, the values represent the percentage of inhibition of: A – Feeding rate (% $I_{FR}$); B – CAT activity (% $I_{CAT}$); C – GSTs activity (% $I_{GSTs}$); D – TBARS levels (% $I_{TBARS}$); E – AChE activity (% $I_{AChE}$), comparatively to the control. Based on the biological responses under study (percent inhibition of different parameters, previously mentioned), ecotoxicity classes were proposed (**Figures 2–4**) to achieve an approach to the ecotoxicological potential for each sampling site (**Figure 5**). Based on the criteria to define the equivalent quality potential, to those presented in the WFD, an estimation of the ecotoxicological potential has been suggested.

For the parameters, feeding rate (FR), and TBARS levels (A and D for each **Figures 2–4**) only two classes of ecotoxicity were defined: non-disturbed (green) and

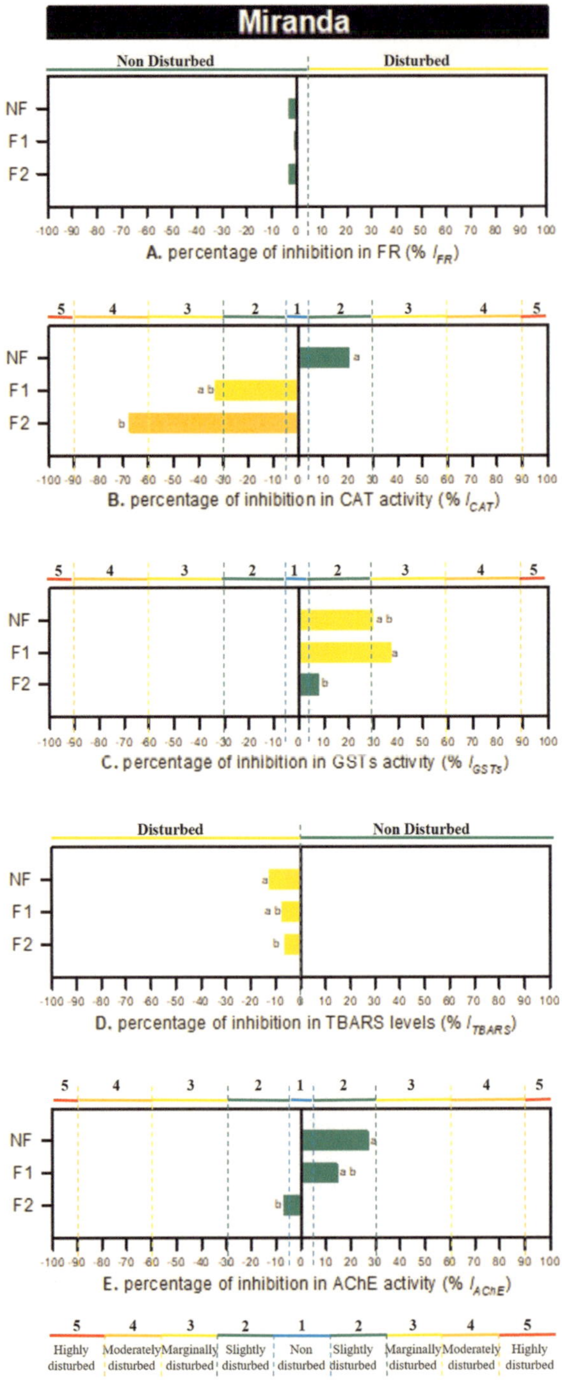


**Figure 2.**

*Proposal of classes of disturbances (defined by colors) and ecotoxicity results for* D. magna, *after exposure to treatments of the natural waters of Pocinho reservoir (NF—Non-filtered water; F1 and F2—Filtered with 1.2 μm and 0.22 μm respectively). The values represent the percentage of inhibition of: A – Feeding rate (% $I_{FR}$); B – CAT activity (% $I_{CAT}$); C – GSTs activity (% $I_{GSTs}$); D – TBARS levels (% $I_{TBARS}$); E – AChE activity (% $I_{AChE}$), comparatively to the control. Different letters (a, b, and c) stand for significant differences between treatments, detected by the Tukey test ($p < 0.05$).*

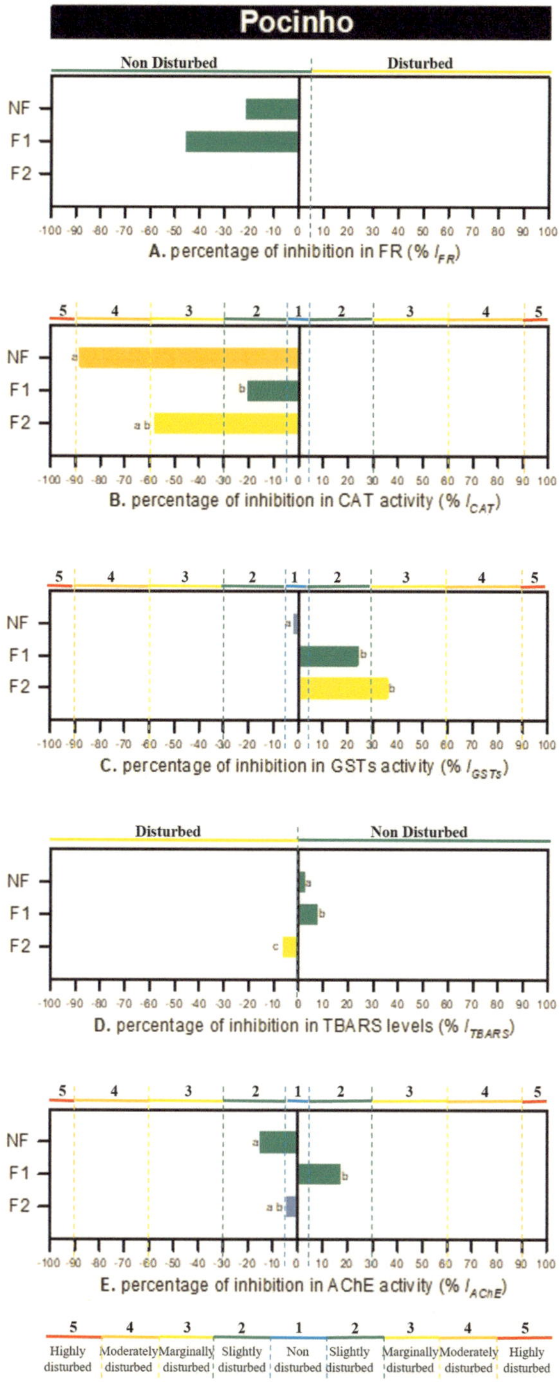


**Figure 3.**
*Proposal of classes of disturbances (defined by colors) and ecotoxicity results for* D. magna, *after exposure to treatments of the natural waters of Miranda reservoir (NF—Non-filtered water; F1 and F2—Filtered with 1.2 μm and 0.22 μm respectively). The values represent the percentage of inhibition of: A – Feeding rate (% IFR); B – CAT activity (% ICAT); C – GSTs activity (% IGSTs); D – TBARS levels (% ITBARS); E – AChE activity (%IAChE), comparatively to the control. Different letters (a and b) stand for significant differences between treatments, detected by the Tukey test ($p < 0.05$).*

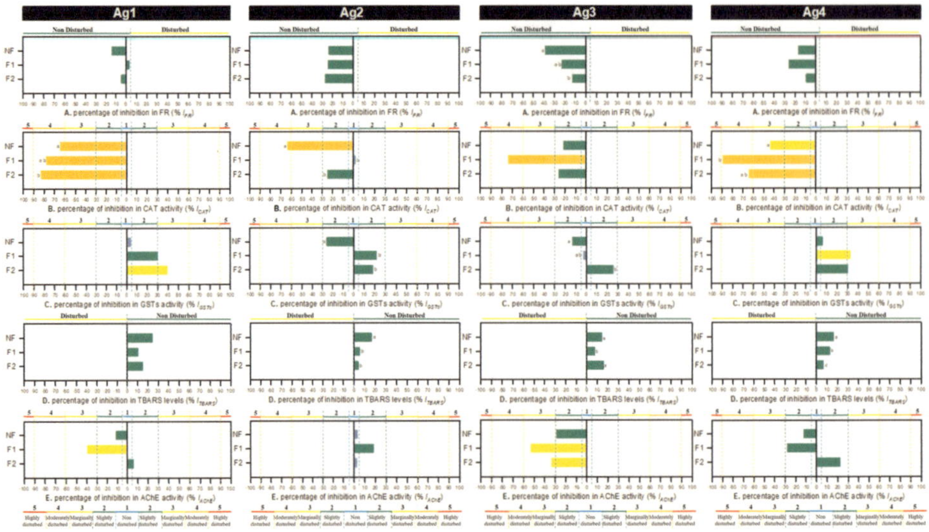


**Figure 4.**
*Proposal of classes of disturbances (defined by colors) and ecotoxicity results for* D. magna*, after exposure to treatments of the natural waters of Aguieira reservoir (NF—Non-filtered water; F1 and F2—Filtered with 1.2 μm and 0.22 μm respectively). The values represent the percentage of inhibition of: A – Feeding rate (% $I_{FR}$); B – CAT activity (% $I_{CAT}$); C – GSTs activity (% $I_{GSTs}$); D – TBARS levels (% $I_{TBARS}$); E – AChE activity (% $I_{AChE}$), comparatively to the control. Different letters (a, b, and c) stand for significant differences between treatments, detected by the Tukey test ($p < 0.05$).*

disturbed (yellow). Different aspects of *Daphnia* biology, as feeding rate is affected by quality (i.e., the number of organic compounds and carbon/nitrogen/phosphate ratio) and quantity of available food [41]. In all reservoirs and sites studied, regarding the parameter of feeding rates (A of **Figures 2–4**), all reservoirs are characterized as not disturbed, since the feeding rates were positive, especially after filtration treatments. However, we draw particular attention to the Ag3 site, where a significant decrease in the percentage of inhibition of the feeding rate, between NF and F2 was observed. This means that, after the filtrations, seston components were removed, including suspended particles, phyto- and zooplanktonic elements, and bacteria, which could be interfering with the feeding capacity of *Daphnia magna*. Lari et al. [41] hypothesized that in addition to physical cues (e.g., concentration and physical properties of seston composition in the water), *Daphnia* detect and uses chemical cues, using their chemosensory system, to locate the most nutritious patches of food in the surrounding environment.

According to [9], lipid peroxidation (LPO) measured as TBARS levels were the most responsive biomarker, in the evaluation of the water quality of the reservoirs under study. If we consider that LPO corresponds to the chain of reactions of oxidative degradation of lipids, resulting in cell damage (e.g., tissue damage), in which free radicals "steal" electrons from the lipids in cell membranes, the distinction of only two classes of ecotoxicities seemed to us to be the most correct and coherent form (D of **Figures 2–4**). We considered that this biological response, that is, the occurrence or not of oxidative damage, results in the generality of the ability of antioxidant defenses to act to prevent, avoid, or neutralize this oxidative damage by free radicals. Organisms can adapt to increasing free radicals (as reactive oxygen or nitrogen species) production by upregulating antioxidant defenses, such as the activities of antioxidant enzymes (e.g., CAT, GSTs, among others) [42]. Failure of

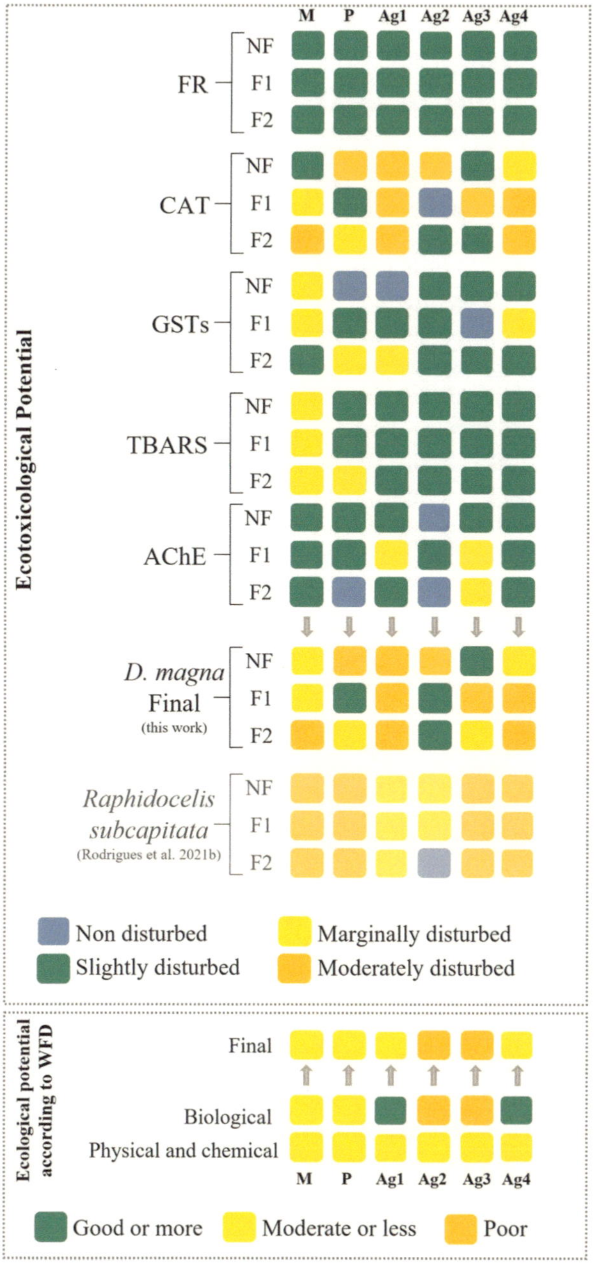


**Figure 5.**
*Ecotoxicological potential of the sampling sites, according to natural water treatments (NF, F1, and F2), ecotoxicity results (defined in previous* ***Figures 2–4****), and ecological potential according to WFD parameters (****Table 1****).*

antioxidant defenses to detoxify excess free radicals production can also lead to significant enzyme inactivation, protein degradation, DNA damage, and lipid peroxidation [42]. In particular, LPO is considered to be a major mechanism, leading to impaired cellular function and alterations in physicochemical properties of cell membranes, which in turn disrupt vital functions of *D. magna*, such as growth, longevity, and reproduction but also feeding behavior. Therefore, an increase in LPO was considered a negative consequence, representing oxidative damage; and a significant decrease will be a positive consequence, that is, the nonoccurrence of oxidative damage, which may be associated with several pathways that avoided, prevented, or neutralized it.

Relatively to the other parameters analyzed (enzymatic activities: B, C, and E of **Figures 2–4**), based on the biological responses under study (percentage of inhibition), five ecotoxicity classes were proposed, as represented in all figures. Based on the criteria to define the equivalent quality potential, to those presented in the WFD, an estimation of the ecotoxicological potential has been suggested. Classes of ecotoxicity have been defined and to facilitate the analysis of global results, different colors were assigned to each class, according to the ecotoxicity degree of the percent inhibition of the parameter under evaluation. For the present work, we consider the following ranges of values (%) and respective ecotoxicity classes: $\leq -5$ to $\geq 5$ (non disturbed—blue); $\leq -5$ to $-30$ and $\geq 5$ to 30 (slightly disturbed—green); $\leq -30$ to $-60$ and $\geq 30$ to 60 (marginally disturbed—yellow); $\leq -60$ to $-90$ and $\geq 60$ to 90 (moderately disturbed—orange); $\leq -90$ and $\geq 90$ (highly disturbed—red). The definition of this range of ecotoxicity classes had as main influences the percentage of effect of 10, 50, and 90% (values with high significance in ecotoxicology), as previously reported by [10]. The range of purposed ranges of ecotoxicity was adjusted, whereby equivalent variations were defined with five ecotoxicity classes, as suggested in the works by [10, 11]. Roig et al. [11] considered an approach to evaluate the ecotoxicological status of rivers (Ebro River watershed, NE Spain), in which the ecotoxicity of pore water has been evaluated in several models organisms, including *D. magna*. Rodrigues et al. [10] and Roig et al. [11] also proposed five classes of ecotoxicity, based on different endpoints, since they evaluated the effects in several aquatic organisms. Roig et al. [11] for *D. magna*, this range was demarcated according to the EC50 values and was expressed as % dilution, for pore water assays, from nontoxic (>100) and highly toxic (<10). Rodrigues et al. [10] defined five ecotoxicity classes for *R. subcapitata*, and this range was defined according to the percent inhibition of yield, from non perturbed ($\geq -10$) and highly perturbed ($<-90$).

Enzymes (e.g., CAT, GSTs, AChE) are proteins that catalyze non-spontaneous chemical reactions in different metabolic pathways, with different physiological functions. CAT is an antioxidant enzymatic defense, GSTs have a dual role in detoxification but also antioxidant defense, and AChE is involved in the neurotransmission process. Enzyme and substrate concentrations influence the reaction rate, altering their activities, which can be significantly inhibited or stimulated [43] after different compound exposure. Antioxidant enzymes (e.g., CAT and GSTs) can be induced by increasing the production of reactive oxygen species (ROS) as a protection mechanism against oxidative stress (adaptation to stress resulting from directly or indirectly generating ROS). In contrast, they can be inhibited when deficiency of the system occurs, inducing a precarious state, making organisms more susceptible to toxic agents (e.g., significant enzyme inactivation or protein degradation by toxicants, or due to the potential attack by excessive concentrations of free radicals) [42, 43]. For example, Hg concentrations (0.08, 0.4, and 2 μg/L) promote perturbations in antioxidant enzymes (e.g. superoxide dismutase; glutathione peroxidase; glutathione reductase; and GSTs) and generate oxidative stress/damage indirectly by binding to antioxidant enzymes containing the thiol group and resulting in depletion of nonenzymatic antioxidant GSH, a scavenger of ROS, for 24 h and 48 h, in neonates and juveniles of *D. magna* [34]. This study corroborates our work, as the quantified Hg concentrations varied between 0.63 and 1.85 μg/L. However, we cannot neglect the mixture of potential compounds present, as well as their interactions and other features of water. Several factors can alter the catalytic activity of enzymes. Altogether, they reflect the current metabolic situations and trigger changes in the inherent characteristics of the enzyme and its interaction to promote or impede enzymatic reactions. Factors such as pH, temperature, effectors, and inhibitors (e.g., chemical compounds dissolved in water) can modify the enzyme concentration and/or conformation but also the substrate concentrations, influencing the reaction rate, and altering its catalytic activity.

AChE is an enzyme involved in the physiological hydrolytic degradation of the neurotransmitter acetylcholine (ACh), in cholinergic synapses and neuromuscular junctions of most organisms [29], and as such, it is indispensable for the normal functioning of the nervous system (neuromuscular transmission) [44]. This biomarker is used as an indicator of neurotoxicity since it results in severe neurotransmission impairment, which leads to ACh accumulation at synaptic clefts, causing nervous overstimulation and eventually death [45]. Changes in normal neurotransmission may have adverse impacts on key functions, such as food consumption, energy metabolism, growth, and reproduction; ultimately, the impairment of neuronal transmission may result in the death of exposed organisms [46]. The US EPA [47] suggests that a significant AChE activity alteration by 20% or more can be considered a clear toxicological effect of stress exposure. However, we agree that the greater the effect on this biomarker, the worse the final consequence, in terms of the aforementioned sub-individual effect (e.g., food consumption, growth, reproduction, and escape from predators). Another factor affecting AChE activity is allosteric control, which can involve stimulation of enzyme action as well as inhibition. Allosteric stimulation and inhibition allow the production of energy and materials by the cell when they are needed and inhibit production when the supply is adequate [48]. The rate of an enzymatic reaction increases with increased substrate concentration, reaching maximum velocity when all active sites of the enzyme molecules are engaged. The cholinergic system plays a major role in the neurotransmission process, and the simultaneous stimulation of nicotinic and muscarinic receptors by ACh may be necessary to synchronize and balance ionic and metabolic events within cells, which are perturbed [49]. Thus, an increase in AChE activity can be associated with perturbations in several metabolic pathways, which can be mediated by ACh. External factors such as food supply, ambient temperature, and water quality (e.g. contaminants mixture) can also alter the activity of cholinesterases [50]. These factors impair the determination of the "normal" activity of ChE and thus hinder the identification of "abnormal" activity, including that caused by anticholinesterases [50].

#### *3.3.1 Ecotoxicity results from the reservoir and its relationship with WFD parameters (physical and chemical elements, chemical analysis, and biological element)*

##### *3.3.1.1 Miranda reservoir*

**Figure 2** corresponds to the proposal of classes of disturbances (defined by colors) and ecotoxicity results for different parameters quantified in *D. magna*, after exposure to treatments of the natural waters of Miranda reservoir. Based on the WFD parameters previously discussed, the feeding rates of *D. magna* exposed to natural waters from Miranda (NF) were not affected (**Table 2**) by the high presence of

| | FR | | | CAT | | | GSTs | | | TBARS | | | AChE | | |
|---|---|---|---|---|---|---|---|---|---|---|---|---|---|---|---|
| | d.f. | F | *p* | d.f. | F | *p* | d.f. | F | *p* | d.f. | F | *p* | d.f. | F | *p* |
| M | 2, 13 | 0.020 | 0.981 | 2, 8 | 10.13 | **0.012** | 2, 8 | 5.554 | **0.043** | 2, 8 | 7.186 | **0.026** | 2, 8 | 7.282 | **0.025** |
| P | 2, 14 | 3.163 | 0.079 | 2, 8 | 5.803 | **0.040** | 2, 8 | 33.136 | **0.001** | 2, 8 | 47.753 | **<0.001** | 2, 8 | 6.777 | **0.029** |
| Ag1 | 2, 14 | 1.522 | 0.258 | 2, 8 | 4.756 | **0.048** | 2, 8 | 5.659 | 0.154 | 2, 8 | 0.870 | 0.466 | 2, 8 | 1.347 | 0.329 |
| Ag2 | 2, 14 | 0.065 | 0.938 | 2, 8 | 21.056 | **0.002** | 2, 8 | 8.066 | **0.020** | 2, 8 | 87.577 | **<0.001** | 2, 8 | 2.425 | 0.169 |
| Ag3 | 2, 14 | 4.791 | **0.030** | 2, 8 | 2.740 | 0.143 | 2, 8 | 6.310 | **0.033** | 2, 8 | 99.326 | **<0.001** | 2, 8 | 1.054 | 0.405 |
| Ag4 | 2, 14 | 2.889 | 0.095 | 2, 8 | 4.756 | **0.048** | 2, 8 | 2.596 | 0.154 | 2, 8 | 408.549 | **<0.001** | 2, 8 | 3.646 | 0.092 |

**Table 2.**
*ANOVA summary table (test differences between natural water treatments – NF, F1, and F2) for the* D. magna *feeding rate (FR), CAT and GSTs activities, TBARS levels, and AChE activity, for Miranda, Pocinho and Aguieira. For each one, the degrees of freedom (d.f.), F statistics, and associated* p*-value was shown. Bold values stand for statistically significant differences. Significant values, (after Tukey test,* $p < 0.05$*), were represented in the figures with different letters (a, b, c).*

cyanobacteria, nor by the levels of total phosphorus, or the concentrations of zinc and mercury, which were recorded above the EQS. Even after the filtration treatments (F1 and F2), these rates were not altered and *D. magna* did not show disturbances in terms of feeding rates (**Figure 2A**; **Table 2**). However, in sub-individual terms (biomarkers), regarding CAT activity, we found that with the application of treatments with filtrations (F1 and F2) there was an increase in oxidative stress (**Table 2**), which resulted in a worse classification class in F2 treatment (moderately disturbed; **Figure 2B**). These results are supported by previous work [8, 9], since higher concentrations of phosphorus can promote an overgrowth of phytoplanktonic organisms [9] and, consequently, the blooms of cyanobacteria as *Microcystis*, which fact confirms our findings, in this work. In turn, these cyanobacterial blooms can result in dangerous levels of toxins such as microcystin-LR toxic to *D. magna* [9, 51–53]. Furthermore, it is important to note that the observed variations in CAT activity may still be associated with mercury and zinc levels above the Eq. [9], which may be more bioavailable for *D. magna* after F2 treatment since only the seston components were removed.

Regarding the activity of GSTs, there was an improvement in the ecotoxicological classification, with the application of the F2 treatment (slightly disturbed; **Figure 2C**). As mentioned in previous studies, changes in the activities of antioxidant enzymes, such as CAT and GSTs, may be associated with the physiological responses of organisms to environmental adaptations, through the influence of phyto and zooplanktonic communities and suspended particles [9]. Chemical analyzes showed very low levels of most quantified contaminants, except for mercury and zinc. On the other hand, as suggested by [54] nutrients seem to be very important in controlling the performance of *D. magna*, and in fact, this corroborates the results observed in the F2 treatment, in the case of GSTs activity.

The results of the TBARS levels showed significant differences between NF and F2 treatments, although some degree of oxidative damage is still observed (**Figure 2D**; **Table 2**). This may indicate that the samples contained some type of disturbing and oxidizing agent, and potentially triggered oxidative stress (previously discussed), with a consequent increase in peroxidative damage (LPO). As mentioned earlier, high amounts of cyanobacteria present in this site, concomitantly with high concentrations of phosphorus and in addition to the high levels of mercury, can be associated with the results of TBARS levels. The accumulation of nutrients (e.g., phosphorus total) in Miranda reservoir can lead to eutrophication causing abnormal growth of the primary producers, which can compromise the quality and balance of the aquatic ecosystem, including the balance between biochemical pathways and physiological functions of organisms, as mentioned by [9, 53].

The results of the AChE activity showed significant differences between NF and F2 in the water treatments, although some degree of neurotoxic alteration was still observed between NF and F2 treatments, an improvement was observed considering the associated ecotoxicological class (**Figure 2E**; **Table 2**). AChE activity stimulation and inhibition allow the production of energy and materials by the cell when they are needed and inhibit production when the supply is adequate [48], and in this study seems to have been affected by the seston components, as can be seen from the changes between NF and F2. A direct relationship between the degree of AChE inhibition and toxicity might not always be expected. The reason for such variability can mainly be attributed to biological differences between species that include AChE sequence differences as well as differences in molecule affinities for the AChE-active site.

*3.3.1.2 Pocinho reservoir*

**Figure 3** corresponds to the proposal of classes of disturbances and ecotoxicity results for different parameters quantified in *D. magna*, after exposure to treatments of the natural waters of Pocinho reservoir. Similar to what was observed for Miranda the feeding rates of *D. magna* exposed to natural waters from Pocinho (NF) were not significantly affected (**Table 2**) by the high presence of cyanobacteria (57.32%), nor by the levels of total phosphorus, or the concentrations of zinc and mercury, which were recorded above the EQS (**Table 1**). Even after the filtration treatments (F1 and F2), these rates were not significantly altered, which shows that in terms of feeding behavior, *D. magna* did not show significant disturbance in terms of feeding rates, despite apparent differences between treatments (**Figure 3A**). However, in sub-individual terms (biomarkers), regarding CAT activity, we found that with the application of treatments with filtrations (F1 and F2), an improvement in the ecotoxicological category was observed between the NF and F1 treatments, but the classification from F1 to F2 worsened again (**Figure 3B**). However, in general, there was an improvement between NF and F2, although not significant (**Table 2**), but sufficient to decrease the ecotoxicological category from moderately disturbed to marginally disturbed. Thus, at this location, we can see a potential negative interference of the various components of seston in CAT activity, which were removed by the F1 treatment, essentially highlight by the percentage of cyanobacteria. In F2, we may have a potential influence of the high concentration of zinc (80.7 μg/L), compared to the other sites and reservoirs (**Table 1**). In the absence of other biological communities, zinc bioavailability may be greater and result in increased toxicity to *D. magna*. Zn toxicity thresholds of *D. magna* can alter by a factor > 10 as a result of ecological interactions and are highly dependent on $P_{total}$ and pH value, with the lowest Zn thresholds found in higher-P and higher-pH waters [55]. However, only a few cases corroborate this finding in this work, and Pocinho does not fit into this perspective. Furthermore, Fettweis et al. [55] evaluate the effects of 25 to 310 μg/L of Zn and pH 7.3 and 7.8 on 21-d daphnid population size and they concluded that the indirect effects of Zn via producer-consumer relationships can outweigh the direct toxic effects. According to the mentioned work, a higher phytoplankton Zn sensitivity at higher pH, affecting food supply to *D. magna*, and an increased algal P content at higher Zn, offering a nutritional benefit to daphnids that counteracts direct Zn toxicity under P limitation [55]. These explanations can help to understand what happened between NF for F1 and F2, not only in sub-individual but also individual responses (feeding rate).

Regarding GSTs activity (**Figure 3C**), a worsening was observed between NF and filtration treatments (F1 and F2). In this case, we can refer to a potential greater bioavailability of metal levels (Zn and Hg or others), in the absence of all seston components, essentially the various biological communities in F2 treatment. These results (**Table 2**) may indicate a potential interference of GSTs, in the antioxidant defense or detoxification of eventually dissolved compounds, which increased along with the water treatments, from NF to F2.

The levels of TBARS (**Figure 3D**; **Table 2**), all previous findings, both in terms of analysis of our results (potential greater bioavailability of dissolved compounds in F2 treatment for *Daphnia*, which is a filtering organism) and by comparison with other studies [9, 55], are reflected in this parameter, since it obtained a worse classification (disturbed), after F2 treatment, comparatively with NF treatment. In fact, Rodrigues et al. [9] had already selected TBARS levels as a relevant parameter in ecotoxicological assessment studies of water quality with *D. magna*. Then, since oxidative damage

(TBARS levels) is an indicator of LPO as a potential consequence of oxidative stress, we can infer that it may be associated with the functional inefficiency of antioxidant enzymes. However, the changes observed in the enzymatic activities of CAT and GSTs support this fact.

The results regarding the AChE activity show an improvement between NF and F2, wherein in this last treatment we achieved the classification of not disturbed. These results indicate that the seston components (present in NF) could be causing some degree of neurotoxic stress, but when organisms were exposed to F2 treatment, these effects were mitigated. In this case, based on EQS values, the altered WFD parameters (pH, $P_{total}$, Zn, and Hg), did not show potential toxicity in terms of neurotransmission.

#### *3.3.1.3 Aguieira reservoir*

**Figure 4** corresponds to the proposal of classes of disturbances (defined by colors) and ecotoxicity results for different parameters quantified in *D. magna*, after exposure to treatments of the natural waters of Aguieira reservoir. Similar to what was observed for the studied Douro River reservoirs (Miranda and Pocinho), the feeding rate parameter, in all treatments demonstrated a classification of not disturbed for all studied locations (**Figure 4A**). Apart from the Ag3 site, no significant differences in this parameter, among the treatments were recorded (**Table 2**). This site presented characteristics that differentiated them from the other Aguieira sites, namely the higher levels of total phosphorus recorded (0.08 mg P/L). Furthermore, higher percentages in terms of cyanobacteria biovolume (10.75%) were observed at this site. In fact, this site was the one that showed the greatest concern in previous studies [8, 10, 12] due to the high nutrient levels, and other metal elements (Zn, Hg), conditions also observed in the present study. In this sense, eutrophication in this reservoir, reported by other authors [9, 10, 12, 31, 40], has been a concern in terms of water quality, and in fact, supports our results. Indeed Ag3 site, with the application of filtration treatments (mainly F2), showed an improvement in feed rates, which reinforces the idea that the presence of several seston components is potential stress inductors for *D. magna* that can be reflected in food behavior.

Contrary to expectable, and about the antioxidant defense biomarker, catalase activity (**Figure 4B**; **Table 2**), a worse classification was observed in the locations Ag1 and Ag4 (in general moderately disturbed), compared to Ag2 and Ag3. In Ag2 and Ag3, a classification of slightly disturbed was observed after filtrations (e.g., F2 treatment). This classification is taking into account the seston components; on the other hand, Ag1 and Ag4 report different scenarios. These sites presented high pH values, and levels of Zn and Hg higher than defined EQS (**Table 1**). In addition, an ecological potential, taking into account biological EQR, the classification was good or more (**Table 1**). For this site, we observe that from NF to F2 there was an increase in oxidative stress, possibly associated with greater availability of dissolved compounds (besides quantified) to *D. magna*, in F2 treatment. In fact, the results of GSTs activity support this potential finding (**Figure 4C**), but only for Ag1. According to Rostern [30], the pH and concentration of inorganic metal ions (e.g., Zn and Hg) are key factors for toxicity in the physiology and behavior of aquatic organisms. In previous studies, metals such as Zn and mainly Hg are known to be inhibitors of enzymes' activities and can disrupt antioxidant defenses [9, 30, 34, 35]. However, some care must be considered when comparing sites in the same reservoir, as we only analyze a tiny part of potential dissolved compounds. Moreover, the biological responses observed in the biomarkers, represent the results of an integrated response to complex mixtures. Furthermore, the land occupation also represents an important source of variation of contaminants input, and consequently biomarkers' response. In addition, land occupation is different in each site and also between reservoirs (**Figure 1**).

Regarding the TBARS levels (**Figure 4D**), despite significant variations between treatments (**Table 2**), in the various sites, the results allowed classifying all treatments as not disturbed. Based on this evidence, we can observe that the antioxidant defenses, despite being altered and indicating potential oxidative stress associated with natural water treatments, in terms of lipid peroxidation, were able to prevent the occurrence of oxidative damage in exposed organisms. In fact, we cannot consider only the antioxidant defenses involved in this work. Other unquantified defenses and metabolic pathways may have acted to neutralize and prevent the occurrence of lipid peroxidation, which could indicate disturbances in cell membranes, which were not observed.

The activity of acetylcholinesterase, despite showing high variations in terms of percentage of inhibition (**Figure 4E**), was not significant between treatments (**Table 2**), for organisms exposed to treatments, with water samples from Aguieira site. The nonoccurrence of significant differences may be associated with high differences between replicates for the same treatment. The different toxicity for organisms exposed to the same conditions, but belonging to different replicates, can be considered if we considered the intraspecific variations [56]. In this sequence, the authors refer that the origin of population, animal body size, and pre-exposure history [e.g., organisms from different cultures, different broods (although meeting the criteria of the assay guidelines)] are realistic variables for zooplankton populations that cause different acute toxicities in *D. magna* [56]. If we consider, for example, the results of the mercury concentrations detected (**Table 1**), potential neurotoxicity would be expected. Mercury is known to be a neurotoxin that causes structural damage to the brain and inhibits enzymes' activities needed for normal neurotransmission [30]. Tsui and Wang [56] suggested that acute Hg toxicities were not simply caused by the different Hg body burdens, and several other mechanisms may operate to result in such a varied Hg toxicity (e.g., reduction of Hg uptake, enhancement of intrinsic tolerance, and increase of antioxidant/detoxification activity). The same authors also measured the metal concentrations in water and living *D. magna* and the results provided useful information to explain whether the apparent tolerance modification was due to a change in metal accumulation and/or to a change in other subtle parameters (e.g., intrinsic tolerance and detoxification activity), which can also alter the neurotoxicity results.

#### *3.3.1.4 Physicochemical and ecological potential vs. ecotoxicological approach*

One of the fundamental using biomarkers in ecological risk assessment is based on their potential ability to anticipate effects at higher levels of biological organization. Due to the different sensitivity between ecotoxicological tools evaluated in this work, and the presence of some confounding factors that could play an important role in the final ecotoxicity evaluation, the result of the final ecotoxicological potential has been calculated as the worst classification of all biochemical parameters quantified in *D. magna* (**Figure 5**). According to this methodology, CAT and AChE activities parameters were the ones that most contributed to the final ecotoxicological potential. Except for Miranda, for the final NF treatment, in which the yellow color resulted from the parameters GSTs activity and TBARS levels. If we consider the totality of the data, we do not observe an improvement in the ecotoxicological potential, with the filtration treatments, with a few exceptions. The only site where this improvement was evident was Ag2, which is corroborated by previous studies conducted in the same site but using the microalga *R. subcapitata* [10]. As was done in the previous study, and due to the different sensitivity between WFD parameters (physicochemical, chemical, and biological elements) individually considered, the final ecological potential according to WFD parameters has been calculated. This approach is more in line with the results of the ecotoxicological potential, since considering the contamination classes of ecotoxicity and respective colors, they present greater similarities (**Figure 5**) with NF treatment (with all components), with the site Ag3 representing an exception. Several reasons were pointed out throughout the manuscript, evidenced by this site, compared to the remaining Aguieira sites reservoir, or even with Miranda and Pocinho. In general, when we considered the sampling points where the ecological potential qualification was bad or poor the similarities between ecologic and ecotoxicological potential were fully agreed upon. The results of the present work allow us to confirm that, when chemical stressors or seston components affect the organism's homeostasis, an ecotoxicological approach, provided by suitable ecotoxicological tools, could detect these changes with accurate sensitivity. In fact, Rodrigues et al. [9] already demonstrated that feeding bioassays and biomarkers (e.g., antioxidant defense and TBARS levels) proved to be useful and reliable tools in the assessment of water quality. Notice that an accurate battery of ecotoxicological tools is a direct measure of organism functional responses, and they could have more impact on the decision-making process than criteria based on concentrations of chemicals or other physical and chemical parameters, as previously demonstrated by [10, 11].

## 4. Conclusions

The results of the current case study corroborate that cost-effective and rapid screening short-term ecotoxicological tools, performed with natural waters, using the model organism *D. magna*, could be useful to complement the determination of the water ecological potential of reservoirs. In the case study of Portuguese reservoirs, ecotoxicological tools evaluated (feeding bioassay and biomarkers) have been performed obtaining good sensitiveness and complementarity between methodologies, in most situations (intra- and inter-reservoirs). Moreover, high coincidences with the ecological potential, recognized following the WFD parameters evaluation, have been found especially when ecosystems' disturbance due to several stressors was observed (e.g., seston components as some phytoplanktonic organisms as cyanobacteria, presence of organic pollutants, and metals). For future comparative studies, we also suggest the evaluation of the water treatments like those performed in this work, mainly the F2 (water filtered through a sterile filter system with a porosity of 0.22 μm), to evaluate seston quality, as it proved to be an important source of stress for *D. magna*. This set of biological responses has presented good concordance with the ecological potential of the reservoirs. These results encourage working further on the applicability of cost-effective ecotoxicity tests and early warning tools for the evaluation of water quality and their integration into the current monitoring programs.

## Acknowledgements

This work was supported by National Funds (through the FCT—Foundation for Science and Technology) and by the European Regional Development Fund (through

COMPETE2020 and PT2020) through the research project ReDEFine (POCI-01-0145-FEDER-029368) and the strategic program UIDB/04423/2020 and UIDP/04423/2020. Sara Rodrigues and Sara Antunes are hired through the Regulamento do Emprego Científico e Tecnológico – RJEC from the Portuguese Foundation for Science and Technology (FCT) program (2020.00464.CEECIND and CEECIND/01756/2017, respectively).

## Conflict of interest

The authors declare no conflict of interest.

## Ethical approval

All applicable international, national, and/or institutional guidelines for the care and use of animals were followed.

## Author details

Sara Rodrigues[1,2]*, Ivo Pinto[1], Sandra Nogueira[1] and Sara C. Antunes[1,2]

1 Faculty of Sciences, Department of Biology, University of Porto, Porto, Portugal

2 CIMAR/CIIMAR, Interdisciplinary Center for Marine and Environmental Research, University of Porto, Matosinhos, Portugal

*Address all correspondence to: sara.rodrigues@fc.up.pt

## References

[1] European Parliament and of the Council. Directive 2013/39/EU of the European Parliament and of the Council of 12 August 2013 amending Directives 2000/60/EC and 2008/105/EC as regards priority substances in the field of water policy. Official Journal of the European Union. 2013;**226**:1-17

[2] Agência Portuguesa do Ambiente. Plano de Gestão de Região Hidrográfica—Região Hidrográfica Do Vouga. Mondego E Lis (Rh4). 2016: 1-169

[3] Santos JI, Vidal T, Gonçalves FJM, et al. Challenges to water quality assessment in Europe—Is there scope for improvement of the current water framework directive bioassessment scheme in rivers? Ecological Indicators. 2021;**121**:107030. DOI: 10.1016/j.ecolind.2020.10703

[4] Arenas-Sánchez A, Rico A, Rivas-Tabares D, et al. Identification of contaminants of concern in the upper Tagus river basin (Central Spain). Part 2: Spatio-temporal analysis and ecological risk assessment. Science and Total Environment. 2019;**667**: 222-233. DOI: 10.1016/j.scitotenv.2019.02.286

[5] Carere M, Antoccia A, Buschini A, et al. An integrated approach for chemical water quality assessment of an urban river stretch through effect-based methods and emerging pollutants analysis with a focus on genotoxicity. Journal of Environmental Management. 2021;**300**:113549. DOI: 10.1016/j.jenvman.2021.113549

[6] De Baat ML, Kraak MHS, Van der Oost R, et al. Effect-based nationwide surface water quality assessment to identify ecotoxicological risks. Water Research. 2019;**159**:434-443. DOI: 10.1016/j.watres.2019.05.040

[7] Escher BI, Stapleton HM, Schymanski EL. Tracking complex mixtures of chemicals in our changing environment. Science. 2020;**367**(80): 388-392. DOI: 10.1126/science.aay6636

[8] Pinto I, Rodrigues S, Lage OM, Antunes SC. Assessment of water quality in Aguieira reservoir: Ecotoxicological tools in addition to the water framework directive. Ecotoxicology and Environmental Safety. 2021;**208**:111583. DOI: 10.1016/j.ecoenv.2020.111583

[9] Rodrigues S, Pinto I, Martins F, et al. Can biochemical endpoints improve the sensitivity of the biomonitoring strategy using bioassays with standard species, for water quality evaluation? Ecotoxicology and Environmental Safety. 2021;**215**:112151. DOI: 10.1016/j.ecoenv.2021.112151

[10] Rodrigues S, Pinto I, Formigo N, Antunes SC. Microalgae growth inhibition-based reservoirs water quality assessment to identify ecotoxicological risks. Water. 2021;**13**:2605. DOI: 10.3390/w13192605

[11] Roig N, Sierra J, Nadal M, et al. Assessment of sediment ecotoxicological status as a complementary tool for the evaluation of surface water quality: The Ebro river basin case study. Science and Total Environment. 2015;**503–504**: 269-278. DOI: 10.1016/j.scitotenv.2014.06.125

[12] Pinto I, Calisto R, Serra CR, et al. Bacterioplankton community as a biological element for reservoirs water quality assessment. Water. 2021;**13**(20): 2836. DOI: 10.3390/w13202836

[13] Pinto I, Rodrigues S, Antunes SC. Assessment of the benthic macroinvertebrate communities in the evaluation of the water quality of Portuguese reservoirs: An experimental approach. Water. 2021;**13**:3391. DOI: 10.3390/w13233391

[14] Rodrigues S, Pinto I, Martins F, et al. An ecotoxicological approach can complement the assessment of natural waters from Portuguese reservoirs? Environemental Science and Pollution Research. 2022:1-15. DOI: 10.1007/s11356-022-19504-4

[15] ISO 10304-1. Water quality—Determination of dissolved anions by liquid chromatography of ions—Part 1: Determination of bromide, chloride, fluoride, nitrate, nitrite, phosphate and sulfate. 2007

[16] ISO 5663. Water quality—Determination of Kjeldahl nitrogen—Method after mineralization with selenium. 1984

[17] ISO 14911. Water quality—determination of dissolved $Li^+$, $Na^+$, $NH_4^+$, $K^+$, $Mn_2^+$, $CA_2^+$, $Mg_2^+$, $Sr_2^+$ and $Ba_2^+$ using ion chromatography—Method for water and waste water. 1999

[18] ISO 17294-2. Water quality—Application of inductively coupled plasma mass spectrometry (ICP-MS)—Part 2: Determination of selected elements including uranium isotopes 31. 2016

[19] INAG I.P. Manual para a avaliação da qualidade Biológica da água em lagos e albufeiras segundo a diretiva quadro da água. In: Protocolo de amostragem e análise para o fitoplâncton; Ministério do Ambiente, do Ordenamento do Território e do Desenvolvimento Regional, Instituto da Água. Lisbon, Portugal: Ministério do Ambiente, do ordenamento do território, e do desenvolvimento regional; 2009. pp. 1-67

[20] ASTM. Standard Practice for Conducting Acute Toxicity Tests with Fishes, Macroinvertebrates, and Amphibians, Reports E 729–80. Philadelphia, USA: American Society for Testing and Materials; 1980

[21] Baird D, Soares A, Girling A, et al. The long-term maintenance of *Daphnia magna* Straus for use ecotoxicity tests: Problems and prospects. In: Lokke H, Tyle H, Bron-Rasmussen F, editors. Proceedings First European Conference on Ecotoxicology. Denmark: Lyngby; 1988. pp. 144-148

[22] Queirós V, Azeiteiro UM, Antunes SC. Feeding inhibition tests as a tool for seston quality evaluation in lentic ecosystems: Salinization impact. Annales de Limnologie—International Journal of Limnology. 2019;**55**(23):1-10. DOI: 10.1051/limn/2019020

[23] Allen Y, Calow P, Baird DJ. A mechanistic model of contaminant-induced feeding inhibition in *Daphnia magna*. Environmental Toxicology and Chemistry. 1995;**14**:1625-1630. DOI: 10.1002/etc.5620140923

[24] Sousa AP, Nunes B. Dangerous connections: Biochemical and behavioral traits in *Daphnia magna* and *Daphnia longispina* exposed to ecologically relevant amounts of paracetamol. Environmental Science and Pollution Research International. 2021;**28**(29): 38792-38808. DOI: 10.1007/s11356-021-13200-5

[25] Bradford MM. A rapid and sensitive method for the quantification of microgram quantities of protein utilizing the principle of protein dye binding.

Analytical Biochemistry. 1976;**72**: 248-254. DOI: 10.1006/abio.1976.9999

[26] Aebi H. Catalase in vitro. Methods in Enzymology. 1984;**6**:105-121. DOI: 10.1016/s0076-6879(84)05016-3

[27] Habig WH, Pabst MJ, Jakoby WB. Glutathione-S-transferases – The first enzymatic step in mercapturic acid formation. The Journal of Biological Chemistry. 1974;**249**(22):7130-7139. DOI: 10.1016/S0021-9258(19)42083-8

[28] Buege JA, Aust SD. Microsomal lipid peroxidation. Methods in Enzymology. 1978;**52**:302-310. DOI: 10.1016/s0076-6879(78)52032-6

[29] Ellman GL, Courtney D, Andres VJ, et al. A new and rapid colorimetric determination of acetylcholinesterase activity. Biochemical Pharmacology. 1961;7:88-95. DOI: 10.1016/0006-2952(61)90145-9

[30] Rostern NT. The effects of some metals in acidified waters on aquatic organisms. Fish & Ocean Opj. 2017;**4**(4): 555645. DOI: 10.19080/OFOAJ.2017.04.555645

[31] Cabecinha E, Cortes R, Alexandre Cabral J, et al. Multi-scale approach using phytoplankton as a first step towards the definition of the ecological status of reservoirs. Ecological Indicators. 2009;**9**:240-255. DOI: 10.1016/j.ecolind.2008.04.006

[32] Kroll SA, Llacer CN, De La Cruz CM, et al. The influence of land use on water quality and macroinvertebrate biotic indices in river within Castilla-La Mancha (Spain). Limnetica. 2009;**28**: 203-214. DOI: 10.23818/limn.28.16

[33] Masindi V, Muedi KL. Environmental contamination by heavy metals. In: HEM S, Aglan RF, editors. Heavy Metals. London: IntechOpen; 2018. DOI: 10.5772/intechopen.76082

[34] Kim H, Kim J-S, Lee Y-M. Changes in activity and transcription of antioxidant enzymes and heat shock protein 90 in the water flea, *Daphnia magna*—Exposed to mercury. Toxicology and Environmental Health Sciences. 2017;**9**(5):300-308

[35] Kim H, Yim B, Bae C, Lee YM. Acute toxicity and antioxidant responses in the water flea *Daphnia magna* to xenobiotics (cadmium, lead, mercury, bisphenolA, and 4-nonylphenol). Toxicology and Environmental Health Sciences. 2017;**9**: 41-49

[36] He ZL, Yang XE, Stoffella PJ. Trace elements in agroecosystems and impacts on the environment. Journal of Trace Elements in Medicine and Biology. 2005; **19**(2–3):125-140. DOI: 10.1016/j.jtemb.2005.02.010

[37] Szatten D, Habel M. Effects of land cover changes on sediment and nutrient balance in the catchment with cascade-dammed waters. Remote Sensor Research. 2020;**12**(20):3414. DOI: 10.3390/rs12203414

[38] Feranec J, Jaffrain G, Soukup T, Hazeu G. Determining changes and flows in European landscapes 1990–2000 using CORINE land cover data. Applied Geography. 2010;**30**:19-35. DOI: 10.1016/j.apgeog.2009.07.003

[39] Bordalo AA, Teixeira R, Wiebe WJ. A water quality index applied to an international shared river basin: The case of the Douro river. Environmental Management. 2006;**38**:910-920. DOI: 10.1007/s00267-004-0037-6

[40] Vasconcelos V, Morais J, Vale M. Microcystins and cyanobacteria trends in a 14 year monitoring of a temperate

eutrophic reservoir (Aguieira, Portugal). Journal of Environmental Monitoring. 2011;**13**:668-672. DOI: 10.1039/c0em00671h

[41] Lari E, Steinkey D, Steinkey RJ, Pyle GG. *Daphnia magna* increase feeding activity in the presence of four amino acids. Journal of Plankton Research. 2018;**40**(5):537-543. DOI: 10.1093/plankt/fby038

[42] Barata C, Varo I, Navarro JC, et al. Antioxidant enzyme activities and lipid peroxidation in the freshwater cladoceran *Daphnia magna* exposed to redox cycling compounds. Comparative Biochemistry and Physiology Part C. 2005;**140**(2):175-186. DOI: 10.1016/j.cca.2005.01.013

[43] Rodrigues S, Antunes SC, Correia AT, Nunes B. Acute and chronic effects of erythromycin exposure on oxidative stress and genotoxicity parameters of *Oncorhynchus mykiss*. Science Total Environment. 2016;**545**: 591-600. DOI: 10.1016/j.scitotenv.2015.10.138

[44] Lionetto MG, Caricato R, Calisi A, et al. Acetylcholinesterase as a biomarker in environmental and occupational medicine: New insights and future perspectives. BioMed Research International. 2013;**321213**:1-8. DOI: 10.1155/2013/321213

[45] Rodrigues S, Antunes SC, Correia AT, Nunes B. Toxicity of erythromycin to *Oncorhynchus mykiss* at different biochemical levels: Detoxification metabolism, energetic balance, and neurological impairment. Environmental Science and Pollution Research. 2019;**26**(1):227-239. DOI: 10.1007/s11356-018-3494-9

[46] Rhee JS, Kim BM, Jeong CB, et al. Effect of pharmaceuticals exposure on acetylcholinesterase (AChE) activity and on the expression of AChE gene in the monogonont rotifer, *Brachionus koreanus*. Comparative Biochemistry and Physiology Part C. 2013;**158**(4):216-224. DOI: 10.1016/j.cbpc.2013.08.005

[47] US EPA. SCE policy issues related to the food quality protection act. Office of pesticide programs science policy on the use of cholinesterase inhibition for risk assessment of organophosphate and carbamate pesticides. Federal register. 1998:63

[48] Britannica. The Editors of Encyclopaedia. "enzyme". Encyclopedia Britannica. 2022, Available from: https://www.britannica.com/science/enzyme [Accessed: April 27, 2022]

[49] Uberti F, Bardelli C, Morsanuto V, et al. Stimulation of the nonneuronal cholinergic system by highly diluted acetylcholine in keratinocytes. Cells, Tissues, Organs. 2017;**203**:215-230. DOI: 10.1159/000451023

[50] Liu HC, Yuan BQ, Li SN. Developing antibodies from cholinesterase derived from prokaryotic expression and testing their feasibility for detecting immunogen content in *Daphnia magna*. Journal of Zhejiang University-SCIENCE B. 2016;**17**(2):110-126

[51] Chen W, Song L, Ou D, Gan N. Chronic toxicity and responses of several important enzymes in *Daphnia magna* on exposure to sublethal microcystin-LR. Environmental Toxicology. 2005;**20**(3): 323-330. DOI: 10.1002/tox.20108

[52] Freitas EC, Pinheiro C, Rocha O, Loureiro S. Can mixtures of cyanotoxins represent a risk to the zooplankton? The case study of *Daphnia magna* Straus exposed to hepatotoxic and neurotoxic cyanobacterial extracts. Harmful Algae.

2014;**31**:143-152. DOI: 10.1016/j.hal.2013.11.004

[53] Wojtal-Frankiewicz A, Bernasińska J, Frankiewicz P, et al. Response of Daphnia's antioxidant systems to spatial heterogeneity in cyanobacteria concentrations in a lowland reservoir. PLoS One. 2014;**9**: e112597. DOI: 10.1371/journal.pone.0112597

[54] Palma P, Ledo L, Alvarenga P. Ecotoxicological endpoints, are they useful tools to support ecological status assessment in strongly modified water bodies? Science and Total Environment. 2016;**541**:119-129. DOI: 10.1016/j.scitotenv.2015.09.014

[55] Fettweis A, De Schamphelaere K, Smolders E. Zinc toxicity to *Daphnia magna* in a two-species microcosm can be predicted from single-species test data: The effects of phosphorus supply and pH. Environmental Toxicology and Chemistry. 2018;**37**(8):2153-2164. DOI: 10.1002/etc.4171

[56] Tsui MTK, Wang WX. Acute toxicity of mercury to *Daphnia magna* under different conditions. Environmental Science & Technology. 2006;**40**:4025-4030. DOI: 10.1021/es052377g

Chapter 3

# Wetlands as the Preferred Roosting and Breeding Site of Sarus Crane, *Grus antigone* (Linneaus, 1758)

*Sarita Rana and Harsh Gulati*

## Abstract

Indian Sarus Crane, *Grus antigone* a world's tallest flying bird. The present study was conducted from November 2019 to October 2020 in the Dhanauri wetland. A total of 71 individuals of Sarus crane were noted. Consisted of 59 adults and 12 juveniles. A study found that the maximum number of Sarus Crane was found to perform roosting (53%) and nesting (15%) activities in wetland areas as compared to other habitats Wetlands perform several ecological functions such as groundwater recharging, maintenance of balance between drought and flooding situations, nutrient cycle, and also conservation of biodiversity. Wetland encroachment will also be detrimental in near future. If the encroachment will remain, then there may be risk of the disappear of wetland areas resulted in imbalance in the nature. The importance of preserving and promoting the ecosystem of wetlands cannot be over emphasized for the sake of agriculture, water conservation and the economy. In the present study it is investigated that human activities are using wetland area for different purposes like, construction of houses, roads, agricultural land and overfishing etc. In view of this finding, it can be suggested that regular monitoring of the bio-parameters and recording the changes, if any, should help maintain the natural vibrancy of the valuable natural assets of marshlands in Uttar Pradesh.

**Keywords:** Sarus crane, wetlands, roosting, breeding, Dhanauri wetland

## 1. Introduction

Indian Sarus Crane, *Grus antigone* a world's tallest flying bird [1–3]. These birds are monogamous and believed to mate for life and suffer the loss of their mates even to the point of starving to death [4]. Due to their habitat destruction, their number is declining gradually, and also listed as vulnerable species as per the IUCN red list of threatened species [5]. It is the only species of breeding crane in India [6]. These birds mainly prefer to live in a habitat like a wetland, marshland that is covered with water during monsoon season, plenty of paddy rice fields, and grassland [7, 8]. Wetlands play a critical role in maintaining many natural cycles and supporting a wide range of biodiversity. They purify and restock our water, and provide the paddy and fishes that feed millions. They serve as a natural sponge against drought and flooding, protect our shorelines and help fight climate change. Wetlands are particularly important providers of all water-related ecosystem services. They regulate water quantity, groundwater recharge, and can contribute to regulating floods and the impacts of storms. Wetlands also help in erosion control and sediment transport, thereby contributing to land formation and increasing resilience to storms. All these ecosystem services improve water security, including security from natural hazards and climate change adaptation. Wetlands are the major habitat for most of the world's waterbirds and key habitat for migratory species. Almost all of the world's waterbirds use wetlands as feeding and breeding grounds. Migratory water birds use wetlands throughout their range which can sometimes literally be from pole to pole. The feeding, breeding and stop-over areas across and between continents that migratory birds depend on requires coordinated wetlands conservation efforts among many nations. Sarus Crane is considered a flagship species that residing a large wetland covered by agricultural land [9]. These birds are omnivorous, and mainly prefer to feed on insects, small fishes, frogs, and some aquatic plants and seeds as well [3, 6]. Sarus cranes generally make association during the pre-monsoon and winter seasons [3, 10]. Among North India, Uttar Pradesh holds the largest population of this species unfortunately intensive data collection is not available for the state. Wetlands perform several ecological functions such as groundwater recharging, maintenance of balance between drought and flooding situations, nutrient cycle, and also conservation of biodiversity [11, 12]. Sarus cranes usually spend most of their time in wetlands to get respite from extensive heat in summers. They congregate into large groups, prior to breeding season in wetland habitat. Bursting with biodiversity, wetlands are a vital means of storing carbon. Wetlands are also tremendously productive ecosystems that provide a myriad of services to society worldwide. Due to the shrinkage of wetlands, they generally prefer to forage on agricultural crops. The present study was carried out to collect information about the preference for wetland habitat by Sarus Crane during breeding and a non-breeding season in Dhanauri wetland, Uttar Pradesh which is one of the favorite sites of these cranes. Continued population declines of many species of aquatic birds indicate the need for additional understanding of how human activities negatively impact water quality and bird populations and what mitigative actions can be taken.

## 2. Materials and methods

### 2.1 Study area

The study was conducted from November 2019 to October 2020 in the Dhanauri wetland. This wetland is located in Thasrana village of Greater Noida, Uttar Pradesh. Dhanauri wetland is located at $28^0 20'12.54"$ N latitudes and $77^0 37'09.77"$ E longitudes covers a total area of around 100 acres surrounded by agricultural land. The wetland is filled with water in monsoon and is also home to the largest number of Sarus cranes. While a large number of migratory birds make this wetland a good avian diversity spot.

### 2.2 Data collection

Before making regular observations, a preliminary survey was carried out to identify the preferred habitat of Sarus Crane and to know their status. Initial information on Sarus inhabited areas was collected through a questionnaire filled by the villagers, farmers, and other respondents of the area. GIS sampling was carried out in order to find their distribution. eTrex Garmin GPS was used to record different coordinates during the study period. Point count methods were made to count the number of the cranes in the study area, by using recorded GPS coordinates we prepared a distribution map. Arc GIS 10.8 in conjugation with Google earth pro was used to locate various sightings of Sarus Crane. Trail cameras (Cuddy back with motion sensors) and a Digital camera (Canon SX70 HS) were used to document the presence or absence of Sarus Cranes at the nest as well as various threat factors behind their vulnerability. Various activities of the birds were observed at the selected sites by monitoring them during the morning and evening hours; 06.00 h–7.30 h and 17.00 h–19.30 h, respectively. Sarus Cranes were recorded in groups and each flock was treated as a separate sample unit, with one GPS point recorded for each.

## 3. Results

A total of 71 individuals of Sarus crane were noted. Consisted of 59 adults and 12 juveniles. During our study, Sarus crane was seen in pairs or family groups for the whole year but during the non-breeding season for the formation of pairs or finding their mates they form a congregation of 60–70 individuals. The present study found that these birds mainly prefer wetland areas (n = 54) while present in congregations during the non-breeding season while selecting their mate (**Table 1** and **Figure 1**). There were only 7 individuals of Sarus cranes has been observed for a long time which was not associated with a large group and they were continuously foraging in agricultural land. Our study also has suggested that the group with very a much smaller number of juveniles as in group 4 already had 2 pairs i.e., 2 males and 2 females were not interested to join the congregation. It has been observed that these birds engage in social displays to facilitate the pairing of unmated birds.

During the study, a maximum number of Sarus Crane was found to perform roosting (53%) and nesting (15%) activities in wetland areas as compared to other habitats (**Figure 2**). However, foraging was strongly preferred in the agricultural landscape, which is likely due to the availability of food in farmlands. It also has been noted that male attracts females to display dance-like movement and also display beak touching behavior as well as flipping up and down while chasing another individual. Dhanauri wetland holds the maximum number of Sarus Crane (**Figures 3–5**).

| Parameters | Group 1 | Group 2 | Group 3 | Group 4 |
|---|---|---|---|---|
| GPS coordinates | 28$^0$20′12.54” N 77$^0$37′09.77″ E | 28$^0$20′14.18” N 77$^0$37′07.04″ E | 28$^0$20′18.10” N 77$^0$37′14.25″ E | 28$^0$20′08.12” N 77$^0$37′03.83″ E |
| Habitat | Wetland, grassland | Wetland | Wetland | Agricultural land |
| Adult | 23 | 10 | 10 | 16 |
| Juvenile | 02 | 03 | 06 | 01 |
| Total (n) | 25 | 13 | 16 | 17 |

**Table 1.**
*Number of individuals of Sarus crane in different groups.*

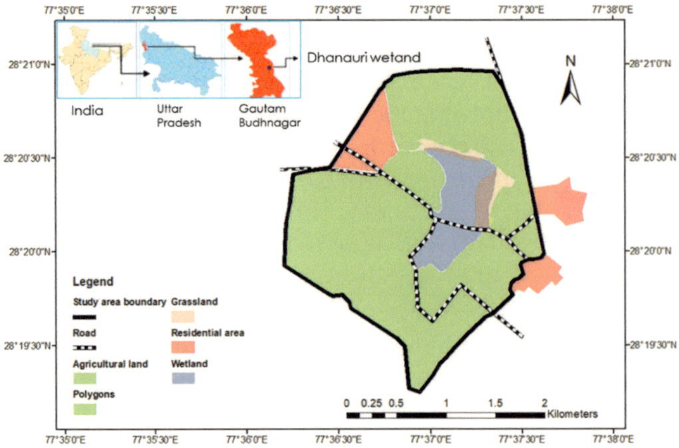


**Figure 1.**
*Map showing 4 different groups noted in Dhanauri wetland.*

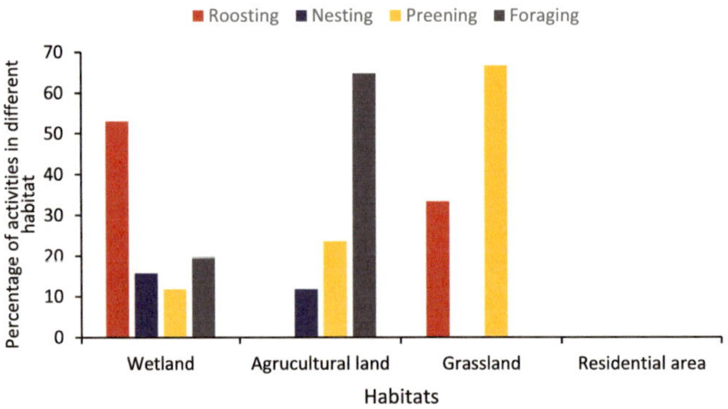


**Figure 2.**
*Various activities performed by Sarus crane in different habitat of Dhanauri wetland.*

**Figure 3.**
*Congregation of Sarus crane in Dhanauri wetland.*

**Figure 4.**
*Roosting of Sarus crane in Dhanauri wetland, Uttar Pradesh.*

**Figure 5.**
*Sarus crane building nest in water.*

## 4. Discussion

The agricultural land around the Dhanauri wetland is also important for Sarus Crane as it frequently uses them to foraging, nest building and breeding. This is largely owing to the fact that this site provides a vast expanse of shallow waters. By way of its long beak, these birds mostly forage in water usually less than 30 cm deep [13]. The ecological suitability of this site in Uttar Pradesh is due to the thriving paddy fields and the absence of the pressure of urbanization. The existing wetlands provide a healthy ecosystem for Sarus cranes and other flora and fauna. Our study revealed that the distribution of Sarus Crane depends upon the suitability of the niche within the habitat. During the non-breeding season, these birds are mostly restricted to the agricultural land, because they spend maximum time foraging for food. Cranes mainly prefer wetland for roosting and nesting and agricultural land for feeding. Instead of building their nests in wetlands, as usual, these birds also move to agricultural land for hiding and to protect their nest and eggs from predators. Due to the ever-increasing urbanization, some other water bodies, that Sarus Cranes may visit, have become unsuitable for nesting during the breeding season. This alarming situation is an eye-opener for the state administration in particular and people in general. The Sarus habitat is outside protected areas, in natural wetlands with low water depth, marshy and fallow areas and agricultural fields. They play a vital role in ecological balance by controlling the population of harmful insects and have significant cultural importance, while also being sociable. Sarus is omnivorous, feeding on fish and insects, as well as roots and plants. Wetlands also help in erosion control and sediment transport, thereby contributing to land formation and increasing resilience to storms. All these ecosystem services improve water security, including security from natural hazards and climate change adaptation. Wetlands are the major habitat for most of the world's waterbirds and key habitat for migratory species. Almost all of the world's waterbirds use wetlands as feeding and breeding grounds. If the protection of this unique habitat of Sarus Cranes receives the priority it deserves, a multidimensional strategy can be drawn to involve all stakeholders. Wetland encroachment will also be detrimental in near future. If the encroachment will remain, then there may be risk of the disappear of wetland areas resulted in imbalance in the nature. Carbon dioxide gas will increase, as a result there will be depleting of ozone layer and it will create several diseases like bronchitis, various skin diseases etc. to living organisms. The homelands of plants, water birds, and animals will be affected. All these are directly or indirectly need for human being. This will degrade the environment near future. In the present study it is investigated that human activities are using wetland area for different purposes like, construction of houses, roads, agricultural land and overfishing etc. The wetland area is going to decrease day by day due to human encroachment. At the grass-root level, village-based education programs and other local developmental programs should be helpful in creating an ambiance for improving the conservation of wetlands [6].

## 5. Conclusion

The importance of preserving and promoting the ecosystem of marshlands cannot be over-emphasized for the sake of agriculture, water conservation, and the economy. The Dhanauri wetland supports a rich diversity of flora and fauna, including a variety of water birds. Due to the ever-increasing pressure of urbanization, a wetland in India is shrinking fast. Besides, several human activities increasingly pollute these precious water bodies. Our study indicates that Sarus Cranes mostly prefer agricultural land near Dhanauri wetland because it is the least disturbed habitat providing protection to their nests and progeny from predators. The direct encouragements to conserve biodiversity as an efficient tool in the delivery of conservation results in a way that also offers developmental benefits to local people. Authors suggest conserving regular monitoring of the wetlands and record the changes if any, should help to maintain the natural vibrancy of the valuable natural assets of wetlands in Uttar Pradesh. The importance of preserving and promoting the ecosystem of wetlands cannot be over emphasized for the sake of agriculture, water conservation and the economy. The wetlands of Uttar Pradesh support a rich diversity of flora and fauna, including a variety of water birds. Due to the ever-increasing pressure of urbanization, wetlands in the study area especially Dhanauri wetland shrinking rapidly. Besides, several human activities increasingly pollute these precious water bodies. Continued population declines of many species of aquatic birds indicate the need for additional understanding of how human activities negatively impact water quality and bird populations and what mitigative actions can be taken. Considering this finding, it is possible to infer that regular monitoring of bio-parameters and documentation of changes, if any, will help in the conserving natural beauty of Uttar Pradesh.

## Acknowledgements

The authors acknowledge the financial support provided by the Council of Scientific & Industrial Research (CSIR), New Delhi. The authors are also thankful to Kurukshetra University for providing the necessary facilities. I acknowledge my gratitude to the Forest & Wild Life Department of Haryana for allowing me to conduct this research in the protected areas of Haryana. However, the authors would also like to thank Amit Kumar and Deepak Kumar Goswami for assisting in data collection during the study.

## Conflict of interest

The authors declare that they do not have any conflict of interest.

**Author details**

Sarita Rana[1*] and Harsh Gulati[2]

1 Department of Zoology, Institute of Integrated and Honors Studies, Kurukshetra University, Kurukshetra, India

2 Department of Zoology, Kurukshetra University, Kurukshetra, Haryana, India

*Address all correspondence to: saritarana20@gmail.com

## References

[1] Ali S, Ripley SD. Handbook of the birds of India and Pakistan. In: Megapodes to Crab Plovers. Vol. 2. Delhi: Oxford University Press; 1980

[2] Meine C, Archibald G. The Cranes: Status Survey and Conservation Action Plan. Gland, Switzerland: IUCN-The World Conservation Union; 1996

[3] Singh HS, Tatu K. A Study on Indian Sarus Crane (*Grus antigone*) in Gujarat State. India: Gujarat Ecological Education and Research (GEER) Foundation, Gandhinagar; 2000

[4] Yaseen M, Saxena R, Dubey S. Population composition, distribution and habitat preference of Indian Sarus crane *Grus antigone* in Chittaurgarh District, southern Rajasthan. Journal of Chemical, Biological and Physical Sciences. 2013;**3**(4):2784-2792

[5] Sundar KSG. Effectiveness of road transects and wetland visits for surveying black-necked storks *Ephippiorhynchus asiaticus* and Sarus crane *Grus antigone* in India. Forktail. 2005;**21**:27-32

[6] Meine C, Archibald GW. Ecology, status and conservation. In: Ellis DH, Gee GE, Mirande CM, editors. Cranes: Their Biology, Husbandry, and Conservation. Washington DC: Department of the Interior, National Biological Service; 1996. pp. 263-292

[7] Archibald GW, Sundar KSG, Barzen J. A review of the three subspecies of sarus crane *Grus antigone*. Journal of Ecological Society. 2003;**16**:5-15

[8] Aryal A. Status and Population of Sarus Crane (*Grus antigone*) in Lowland of West-Central Region of Nepal. Report submitted to the. Nepal: Biodiversity Research and Training Forum; 2004. Available from: http://himalaya.socanth.cam.ac.uk/collections/inskipp/2004_008.pdf

[9] JKK. Save UP wetlands to save flagship species Sarus and in turn wetland biodiversity under the umbrella. In: Water and Biodiversity. Lucknow: Uttar Pradesh State Biodiversity Board; 2013. pp. 17-29

[10] Sundar KSG, Chaudhary BC. Impact of Land-Use Changes on the Ecology and Habitat of the Sarus Crane (*Grus antigone*) in the Indo-Gangetic Flood Plains. Part II. Uttar Pradesh: Wildlife Institute of India; 2008

[11] Bassi N, Kumar MD, Sharma A, Pardha-Saradhi P. Status of wetlands in India: A review of extent, ecosystem benefits, threats and management strategies. Journal of Hydrology: Regional Studies. 2014;**2**:1-19

[12] Sinha SK. Sarus crane conservation project. 2019. Available from: https://www.wti.org.in/projects/sarus-crane-conservation-project.

[13] Sunder KS, Kaur J, Choudhury BC. Distribution, demography and conservation status of the Indian Sarus crane (*Grus antigone*) in India. Journal of Bombay Natural History Society. 2000;**97**(3):319-339

Section 2

# Monitoring Aspects

Chapter 4

# Pollutants of Emerging Concern in Urban Wastewater Impacted Aquatic Environments and Management Recommendations

*Nqobizitha Siziba and Emmanuel Tapiwa Sero*


## Abstract

Contamination of aquatic environments by pollutants of emerging concern (PEC) creates new public health and environmental threats. Over the years, Africa has struggled to adequately treat wastewater before discharged into the environment. The situation is expected to be worsened by the more challenging to treat PEC like pharmaceuticals, endocrine disrupters, microplastics, surfactants, cyanotoxins, radioactive and flame retardants. Generally, the pollution of aquatic environments will have serious negative impacts on organisms that depend on the affected sources. Elsewhere, a number of research studies have reported the occurrence of these pollutants and in some cases exceeding the recommended levels. However, in Africa and other developing countries, a few studies have focused on PEC in aquatic resources. In this review, discussions are centered on the: (i) occurrence of PEC in African aquatic environments, (ii) potential risks to aquatic ecosystems and human health, and (iii) current chemical and biological monitoring techniques. There is need to include PEC in the research studies and routine environmental monitoring programmes particularly before the urban wastewater is discharged into the environment. Passive biomonitoring through using biomarkers like oxidative stress proteins and gonadal histopathology may be more informative and cheaper way of monitoring PEC than chemical analysis.


**Keywords:** pollutants of emerging contaminants, urban wastewater, aquatic ecosystems, biomonitoring

## 1. Introduction

Aquatic ecosystems are core to life as aquatic organisms, terrestrial animals and human beings depend on these fragile environments for water. However, aquatic ecosystems have been negatively affected by a number of anthropogenic factors like damming, overexploitation, agriculture and discharge of poorly treated wastewater. Among these factors, the discharge of poorly treated wastewater from urban settlements is one of the key challenges that continue to degrade the aquatic ecosystems in Africa [1, 2]. The urban wastewater-induced degradation of aquatic ecosystems may be attributed to the slackening development of infrastructure meant for effluent management. To date, most of the water quality studies on aquatic ecosystems in developing countries have largely focused on physicochemical parameters such as heavy metals, biological oxygen demand, dissolved oxygen, suspended solids, chemical oxygen demand etc., whilst negligible attention has been given to pollutants of emerging concern (PEC).

Globally, PEC is increasingly becoming an ecological and public health threat as most of wastewater treatment plants were not designed to extract and treat these. PEC include both synthetic and naturally occurring compounds that are not normally monitored within the aquatic environments but have been recognized as having adverse ecological and health effects. These pollutants have the ability to affect humans and aquatic animals' resident in affected ecosystems even at low doses. The realization of the presence of these pollutants has raised research interests to address source pathways, their fate, transformations and impact on life. A study by [3] grouped the PEC into 11 major types: personal care products, pharmaceuticals, industrial chemicals, polycyclic aromatic hydrocarbons, volatile organic compounds, pesticides, mycotoxins, cyanotoxins, radioactives, microplastics and particulate organic matter.

In Africa, urbanization has been linked to the pollution of downstream water bodies, for example, the downstream pollution of Lake Chivero [1, 4] and Khami Dams [2, 5, 6] in Zimbabwe. This has been attributed to the discharge of poorly treated urban wastewater [7–9]. Therefore, in developing countries, water resources in urbanising catchments are likely to be the hotspots of PEC. Africa has been swiftly urbanizing over the last decades with its urban population having been predicted to triple over 40 years, from 395 million people in 2010 to 1.339 billion in 2050 [10]. However, it is important to note that swift urban expansion in Africa is coupled with slackened sewage infrastructural development resulting in the discharge of poorly treated sewage into the environment. Furthermore, the current sewage treatment technologies that are in use in Africa do not factor in the treatment of PEC. Owing to the pressure linked to the discharge of poorly treated urban wastewater into the environment and a general lack of awareness, the likelihood of pollution by PEC is great. Therefore, there is a need for research and monitoring programmes to focus on PEC, particularly on the urbanizing catchments.

## 2. Pollutants of emerging concern in developing countries

### 2.1 Pharmaceuticals

Pharmaceuticals are largely synthetic organic compounds used in alleviating pain or as antimicrobials, antivirals and contraceptives. The use of pharmaceuticals by humans is followed by excretion of the residual traces of these drugs through urine or faecal matter into the environment. Therefore, with the increasing urban population coupled with the dilapidating wastewater treatment infrastructure, it is expected that large quantities of these pharmaceuticals are discharged into the environment. Elsewhere, the availability of data on contamination of water resources by pharmaceuticals has been increasing [11–14]. In Africa, some studies on pharmaceuticals have been conducted in South Africa [15–17], Kenya [18–20] and Nigeria [3, 21, 22]. This review will focus on antibiotics and contraceptive drugs due to their wider use in municipal areas of developing countries.

In Zimbabwe, the contraceptive method mix is dominated by the pill with more than two-thirds of women using this hormonal method [23]. Estrone was detected in the range of 0.90 to 4.43 ng/L in raw water samples of the Vaal river [24]. The Vaal River drains most of the wastewater from the metropolitan City of Johannesburg and has been termed one of Africa's workhorse. Presence of these contraceptives in aquatic ecosystems has been linked with diverse negative effects that include intersex organisms [25, 26], abnormal secondary sex characteristics [27–29], reduced fecundity [30] and changes in population sex ratios [31]. Most of these effects have dire consequences on the populations of the affected organisms. Wastewater discharged from Bulawayo; Zimbabwe has been reported to be the major source of contraceptive contamination in affected water bodies [6]. In this city, the highest oestrogenic effects were reported in Umguza Dam [6], one of the most polluted dams in Bulawayo [2]. A study by Teta and others [6] reported feminization of male fish in Umguza Dam, a broader threat to aquatic life. On the other hand, Bacterial Antibiotic Resistance (BAMR) is now a global concern that is reversing the progress that had been made in containing bacterial infections [11, 32, 33]. Misuse and overuse of antimicrobials is the main contributing factor to the evolution of antimicrobial resistance (AMR). Elsewhere, antimicrobial-resistant bacteria have been found in the environment with drinking water reported to be the main transmission route of these pathogens to human beings [34]. BAMR is common in areas with extensive use of antibiotics [35] and thus urban wastewater that is discharged into the environment is expected to be rich in pharmaceutical residuals including antibiotics [36–38]. Although antibiotics have been vital in improving human and animal health, these drugs find their way into the aquatic environment largely because of their frequent and unregulated use plus lack of capacity by most urban wastewater treatment plants to remove the antibiotics during wastewater treatment [35, 36]. The antibiotic residues that find their way into the environment exert selective pressure on bacteria leading to the evolution of antibiotic resistance [35, 39].

In Zimbabwe, strains which caused outbreaks of both cholera and typhoid were reported to be drug-resistant with patients showing poor response to commonly used drugs [40–42]. The heavily polluted Lake Chivero, the portable water source for the capital city of Zimbabwe, Harare, and underground water contamination has been cited as some of the possible sources of pathogens causing the recurring outbreaks of diarrheal diseases including cholera and typhoid in the capital. Both chromatography and spectroscopy methods have been used to identify known pharmaceuticals in water [43, 44]. However, the drug sector is developing very fast thus it is always a challenge to come up with a list of chemical compounds to be included for analysis. This makes chemical analysis process much of a daunting task. Furthermore, the pharmaceuticals exist in water below the detection limits of most analytical equipment but at the same time have detrimental effects on resident organisms. Moreover, chemical analyses do not reveal biological effects of pharmaceuticals. Consequently, biomonitoring approaches have been preferred over chemical analyses to assess the effects of pharmaceuticals on organisms. Biomonitoring involves the systematic measurement of the effects of pollutants by focusing on structural, physiological and genetic changes in living organisms as a response to the exposure.

### 2.2 Microplastics

Collectively African countries are estimated to be the second largest contributors of plastic waste to rivers. The plastic waste eventually ends up in oceans [45]. Plastic industry remains core to modern economies and human development. In the year 2016, about 335 million tonnes of plastics were produced [46] and this figure continues to increase in proportion to the increasing human population. The level of microplastic pollution is predicted to be higher in developing countries due to lack of proper waste management facilities which may cause large quantities of plastics to end up in the environment [47]. Plastics can be classified into six types: polyethylene (PE), polypropylene (PP), polyamide (PA), polyvinyl chloride (PVC), polystyrene (PS), polyurethane (PUR), and polyethylene terephthalate (PET) [48–50]. Plastics can also be grouped by their physical structure into five groups: fragments, micro-pellets, fibres, films and foam [51]. Micro pellets originate from different sources like washing powders and paints, fibres derived from synthetic textiles, foam cushioning material and fragments a result of breakdown of items such as plastic bottles and packaging materials [52].

Most plastics degrade into smaller particles over time through mechanical forces, thermo-degradation, photolysis, thermo-oxidation and biodegradation processes [53]. Therefore, microplastics (< 5mm) might arise through degradation or directly from a range of products including washing powder, rinse-off cosmetics and personal care [54, 55]. It is these smaller microplastics (MPs) that are observed to be dominating in the aquatic food webs [56] but with biological effects not well understood. Generally, in Africa, the extent of MPs within the inland waters remains largely unreported. Some of the few studies in South Africa [57], Botswana [58], Nigeria [59], and Kenya [60] have observed the microplastics in gut contents of fish. Analytical techniques that have been used to assess microplastics include the basic light compound microscope, Fourier Transform Infrared Spectroscopy (FTIR), Rahman Spectroscopy, and pyrolysis followed by GC/MS.

### 2.3 Cyanotoxins

Cyanobacteria also known as blue-green algae are part of aquatic algae that are known for producing biotoxins called cyanotoxins. Nutrient enrichment of aquatic systems largely by the discharge of poorly treated urban wastewater [1, 5], agricultural and industrial runoff has increased the proliferation of these harmful algae. The cyanotoxins that are normally produced after cell lysis following the collapse of the algal blooms have hepatotoxic, neurotoxic, carcinogenic and teratotoxic, cytotoxic and dermatotoxic effects [61–63]. The cyanobacteria genera that have been observed to be responsible for the formation of toxic blooms in aquatic systems include *Microcystis*, *Cylindrospermopsis*, *Anabaena*, *Aphanizomenon* and *Planktothrix*. The main cyanotoxins that are produced by these cyanobacteria are microcystins, cylindrospermopsin, anatoxin-a and saxitoxins. According to [64], hepatotoxic microcystins are the most widespread class of cyanotoxins and are widely used as indicators of the presence of cyanotoxins in aquatic systems.

Serious chronic human and animal health problems, and in some cases mortalities have been linked to cyanotoxin poisoning [65, 66]. According to [67], 1.0 μg/L (0.001 mg/L) is the recommended level for microcystin in drinking water whereas 2 000 *Microcystis* cells/mL have been recommended as the upper limit of cyanobacteria in drinking water for animals [68]. Research on cyanotoxins is still limited in Africa with a few studies having been done in Zimbabwe [4, 69] and South Africa [70, 71]. Currently, cyanotoxins are being implicated in the deaths of elephants in Botswana, South Africa, Zimbabwe [72] and fish in Zimbabwe [1]. In Harare, Zimbabwe cyanotoxin poisoning has been linked to the increase in gastroenteritis [73, 74], liver cancer [75] and the death of fish in Lake Chivero [1].

The cyanotoxins have been detected in nutrient-rich systems, particularly those systems that receive poorly treated urban wastewater. The microcystin concentrations ranged between 3.67 to 86.08 mg/L in hypertrophic Hartbeespoort Dam, South Africa and between 0.1 to 49.41 mg/L in Kruger National Park [76]. In a heavily polluted Lake Chivero in Zimbabwe, microcystin concentrations ranged between 18.02 to 22.48 μg/L [75]. High levels of microcystins, ranging from 0.58 to 2.65 $\mu g\ L^{-1}$, were detected in Lake Tana, Ethiopia [77]. The assessment of microcystin in a major drinking water source, Legedadi Reservoir, of Addis Ababa, Ethiopia recorded levels ranging between 61.63 and 453.89 $\mu g\ L^{-1}$ [78]. The few studies done in Maputo and Gaza provinces indicated the occurrence of microcystin-LR, ranging from 6.83 to 7.78 $\mu g \cdot L^{-1}$ [79]. Therefore, levels of cyanotoxins in eutrophic aquatic resources in Africa, such as those that receive poorly treated wastewater, may be above the WHO guideline value of 1.0 $\mu g\ L^{-1}$ and thus pointing towards high risks to public and environmental health. We suggest that quantitative cyanotoxin measurements be included in the water quality monitoring programs as guiding precautionary actions to mitigate the risks to public health and biodiversity.

### 2.4 Surfactants

Globally, the market size of surfactants is about US$42.1 billion, and it is expected to rise to US$52.4 billion by 2025 [80]. Surfactants are utilised in the production of detergents, textiles, paints, polymers, pharmaceuticals, pesticides, paper, personal care products and in mining for the extraction of minerals through flotation. Surfactants are amphipathic molecules possessing both hydrophilic heads and hydrophobic tails [81]. Depending on the type of charge on the hydrophilic head, surfactants can be categorized into five groups: anionic, cationic, non-ionic, semi-polar and amphoteric. Cationic surfactants possess a positive hydrophilic group, while anionic surfactants contain negatively charged hydrophilic functional groups [82]. The non-ionic surfactants (TAS) possess a non-ionized hydrophilic group(s), while the charge on the hydrophilic sites of amphoteric surfactants changes as a function of pH [83, 84]. The anionic, cationic, and amphoteric surfactants constitute 65% of surfactants on the global market size.

Owing to the extensive use of surfactants in urban areas, these molecules are discharged with wastewater and end up contaminating the receiving water bodies. Some of these water bodies like Lake Chivero and Darwendale in Zimbabwe also serve as portable water sources for the municipalities thereby raising public and environmental health concerns. Anionic surfactants like the linear alkylbenzene sulfonates (LAS) have been associated with various ecotoxicological effects on aquatic/terrestrial ecosystems [82, 85] and humans [83]. Surfactants are also known to reduce the resistance of aquatic biota to environmental stresses affecting reproduction and growth processes [86, 87]. Surfactants also increase the solubility of most contaminants thus increasing their toxicity in aquatic environments [88, 89]. The enormous adverse effects of surfactants on health and the environment necessitates the need to include them in environmental monitoring programmes.

## 3. Monitoring of PEC

### 3.1 Chemical analysis and biomonitoring techniques

Monitoring of PEC is critical for conservation, guiding remediation efforts as well as biodiversity and human life protection from the adverse effects of these. This is highly challenging for several reasons: PEC are very diverse and together with their transformed products and their variations will increase as nations push the innovation agenda aimed at coming up with alternatives. Therefore, the available specialised equipment by far does not cover the full range of PEC. Some of these methods are either still being developed or are yet to be adapted to our systems. Furthermore, PEC affects life at low concentrations and there is a need for equipment to have low detection limits to allow for proper risk assessment. Currently, Africa and other developing countries are lagging behind with respect to the required state-of-the-art equipment to competitively monitor the PEC in aquatic bodies. The reasons being that of poor awareness and consequently low budget prioritisations.

In contrast to the challenging and expensive chemical analysis methods, biomonitoring technologies provide a relatively cheaper and more integrated method for monitoring the PEC. Biomonitoring in aquatic ecosystems is the detection of substances or their effects on organisms, compared to analyzing chemical pollutants in water samples. Biomonitoring follows two major approaches: the bioindicator and biomarker methods. The biological indicator method uses the presence or absence of organism(s) to indicate the level of certain critical factors including pollutants [90]. A biomarker is a naturally occurring biological molecule, gene, or structural characteristic expressed as a response to the exposure of an organism(s) or cell(s) to a particular pathological or physiological process, disease or in this case pollutants. Biomarkers are detectable biochemical and tissue-level changes in response to exposure to pollutants. Between these two approaches, it is the biomarker approach that is increasingly being supported for use in monitoring PEC in aquatic systems [91–93]. Biomonitoring through the use of biomarkers can detect the exposure, the effect, or reveal susceptibility. Biomarkers are the best approach to identifying an early response to contaminants [94] and are much more sensitive to identifying organism stress than the community responses. They are normally the key pillar of bioindication method [95]. Using fish and aquatic snails, a number of studies have been conducted in Zimbabwe to assess the biomarkers that are expressed in polluted aquatic ecosystems. These studies have shown the expression of antioxidant enzymes [96–98], histological pathology of gills and liver [99], reduced fecundity and feminization [6] in polluted aquatic ecosystems. These studies conducted in Zimbabwe are providing evidence that biomarkers might provide an alternative to the chemical analysis methods to monitor PEC in aquatic systems of developing countries such as Zimbabwe.

## 4. Conclusions and outlook

This review shows that in Africa, research and monitoring programmes on PEC are still very limited. The fact that PEC is still missing in most water quality and aquatic ecosystems monitoring and research programmes is an indication that awareness of these is still limited and thus they are less prioritised in various initiatives. The few published studies are pointing out that water bodies receiving urban wastewater are likely to be the hotspots of PEC, but the effects are still poorly understood. Chemical analyses, using very expensive equipment that might not be easily accessible in developing countries, are still the main means of generating information on PEC. The few studies that have been conducted in developing countries are showing the presence of PEC and in some cases exceeding the recommended limits. It was highlighted that the other limitation of using the chemical analysis methods stems from the equipment available on the market having poor sensitivity resulting in failure to detect PEC at low concentrations. Therefore, biomonitoring using biomarkers is suggested as an alternative. Although still in its infancy, biomonitoring using biomarkers will provide a cheaper and more integrated way of monitoring PEC in aquatic systems. Therefore, there is a need for more studies aimed at assessing the feasibility of using biomarkers for monitoring PEC in aquatic ecosystems, particularly those that receive urban wastewater. This paper further recommends the need to: (i) improve awareness of the existence of PEC, (ii) assess the effectiveness of the current wastewater treatment plants in the removal of PEC, (iii) include PEC in the routine monitoring programmes by municipalities, (iv) governments and other funding agencies should fund research on PEC to address knowledge gaps.

## Conflict of interest

The authors declare no conflict of interest.

## Author details

Nqobizitha Siziba* and Emmanuel Tapiwa Sero
Biological Sciences Department, Chinhoyi University of Technology, School of Natural Sciences and Mathematics, Chinhoyi, Zimbabwe

*Address all correspondence to: nqobanie@yahoo.co.uk

## References

[1] Moyo NAG. Lake Chivero: A Polluted Lake. Harare, Zimbabwe: University of Zimbabwe Publishers; 1997

[2] Siziba N, Mwedzi T, Muisa N. Assessment of nutrient enrichment and heavy metal pollution of headwater streams of Bulawayo, Zimbabwe. Physics Chemical Earth. 2021;**122**:102912

[3] Egbuna C, Amadi CN, Patrick-Iwuanyanwu KC, Ezzat SM, Awuchi CG, Ugonwa PO, et al. Emerging pollutants in Nigeria: A systematic review. Environmental Toxicology and Pharmacology. 2021;**85**:103638

[4] Mhlanga L, Day J, Cronberg G, Chimbari M, Siziba N, Annadotter H. Cyanobacteria and cyanotoxins in the source water from Lake Chivero, Harare, Zimbabwe, and the presence of cyanotoxins in drinking water. African Journal of Aquatic Science. 2006;**31**:165-173

[5] Siziba N. Effects of damming on the ecological condition of urban wastewater polluted rivers. Ecological Engineering. 2017;**102**:234-239

[6] Teta C, Holbech BF, Norrgren L, Naik YS. Occurrence of oestrogenic pollutants and widespread feminisation of male tilapia in peri-urban dams in Bulawayo, Zimbabwe. African Journal of Aquatic Science. 2018;**43**:17-26

[7] Nhapi I, Hoko Z, Siebel MA, Gijzen HJ. Assessment of the major water and nutrient flows in the Chivero catchment area, Zimbabwe. Physical and Chemical Earth. 2002;**27**:783-792

[8] Nhapi I, Tirivarombo S. Sewage discharges and nutrient levels in Marimba River, Zimbabwe. Watermark. 2004;**30**:107-113

[9] Nhapi I. The water situation in Harare, Zimbabwe: A policy and management problem. Water Policy. 2009;**11**:221-235

[10] Güneralp B, Lwasa S, Masundire H, Parnell S, Seto KC. Urbanization in Africa: Challenges and opportunities for conservation. Environmental Research Letters. 2018;**13**

[11] Kümmerer K. The presence of pharmaceuticals in the environment due to human use - present knowledge and future challenges. Journal of Environmental Management. 2009;**90**:2354-2366

[12] Deblonde T, Cossu-Leguille C, Hartemann P. Emerging pollutants in wastewater: A review of the literature. International Journal of Hygiene Environmental Health. 2011;**214**:442-448

[13] Caban M, Stepnowski P. How to decrease pharmaceuticals in the environment? A review. Environmental Chemical Letters. 2021;**19**:3115-3118

[14] Maculewicz J, Kowalska D, Świacka K, Toński M, Stepnowski P, Białk-Bielińska A, et al. Transformation products of pharmaceuticals in the environment: Their fate, (eco)toxicity and bioaccumulation potential. Science Total Environment. 2022;**802**

[15] Matongo S, Birungi G, Moodley B, Ndungu P. Pharmaceutical residues in water and sediment of Msunduzi River, KwaZulu-Natal, South Africa. Chemosphere. 2015;**134**:133-140

[16] Ngqwala NP, Muchesa P. Occurrence of pharmaceuticals in aquatic environments: A review and potential impacts in South Africa. South African Journal of Science. 2020;**2020**:116

[17] Ojemaye CY, Petrik L. Pharmaceuticals and Personal Care Products in the Marine Environment Around False Bay, Cape Town, South Africa: Occurrence and Risk-Assessment Study. Environmental Toxicoloy Chemistry. 2022;**41**:614-634

[18] K'oreje KO, Kandie FJ, Vergeynst L, Abira MA, Van Langenhove H, Okoth M, et al. Occurrence, fate and removal of pharmaceuticals, personal care products and pesticides in wastewater stabilization ponds and receiving rivers in the Nzoia Basin, Kenya. Science Total Environment. 2018;**637-638**:336-348

[19] Kandie FJ, Krauss M, Beckers LM, Massei R, Fillinger U, Becker J, et al. Occurrence and risk assessment of organic micropollutants in freshwater systems within the Lake Victoria South Basin, Kenya. Science Total Environment. 2020;**714**:136748

[20] Muriuki CW, Home PG, Raude JM, Ngumba EK, Munala GK, Kairigo PK, et al. Occurrence, distribution, and risk assessment of pharmerciuticals in wastewater and open surface drains of peri-urban areas: Case study of Juja town, Kenya. Environmental Pollution. 2020;**267**:115503

[21] Ogunbanwo OM, Kay P, Boxall AB, Wilkinson J, Sinclair CJ, Shabi RA, et al. High concentrations of pharmaceuticals in a Nigerian river catchment. Environmental Toxicology and Chemistry. 2022;**41**:551-558

[22] Ilechukwu I, Okonkwo CJ, Olusina TA, Mpock JA, Ilechukwu C. Occurrence and risk assessment of selected pharmaceuticals in water and sediments of Usuma Dam, Abuja, Nigeria. International Journal of Environmental Analytical Chemistry. 2021;**3**:1-13

[23] Sambisa W, Curtis SL. Contraceptive Use Dynamics in Zimbabwe: Postpartum Contraceptive Behaviour. Maryland, USA: Macro International Inc.; 1997

[24] Mnguni SB, Schoeman C, Marais SS, Cukrowska E, Chimuka L. Determination of oestrogen hormones in raw and treated water samples by reverse phase ultra-fast liquid chromatography mass spectrometry – A case study in Johannesburg south, South Africa. Water. 2018;**44**:111-117

[25] Orlando EF, Guillette LJ. Sexual dimorphic responses in wildlife exposed to endocrine disrupting chemicals. Environmental Research. 2007;**104**:163-173

[26] Geraudie P, Gerbron M, Minier C. Endocrine disruption effects in male and intersex roach (Rutilus rutilus, L.) from French rivers: An integrative approach based on subcellular to individual responses. Composite Biochemical Physiology. 2017;**211**:29-36

[27] Frankel TE, Meyer MT, Orlando EF. Aqueous exposure to the progestin, levonorgestrel, alters anal fin development and reproductive behavior in the eastern mosquitofish (Gambusia holbrooki). Genetics Composite Endocrinology. 2016;**234**:161-169

[28] Wojnarowski K, Podobiński P, Cholewińska P, Smoliński J, Dorobisz K. Impact of estrogens present in environment on health and welfare of animals. Animals. 2021;**11**:1-16

[29] Tomkins P, Saaristo M, Allinson M, Wong BBM. Exposure to an agricultural contaminant, 17β-trenbolone, impairs female mate choice in a freshwater fish. Aquatic Toxicology. 2016;**170**:365-370

[30] Schwindt AR, Winkelman DL, Keteles K, Murphy M, Vajda AM. An

environmental oestrogen disrupts fish population dynamics through direct and transgenerational effects on survival and fecundity. Journal of Applied Ecology. 2014;**51**:582-591

[31] Decourten BM, Brander SM. Combined effects of increased temperature and endocrine disrupting pollutants on sex determination, survival, and development across generations. Scientific Reports. 2017;7:1-9

[32] Bhullar K, Waglechner N, Pawlowski A, Koteva K, Banks ED, Johnston MD, et al. Antibiotic resistance is prevalent in an isolated cave microbiome. PLoS One. 2012;7:1-11

[33] Ayandiran TA, Ayandele AA, Dahunsi SO, Ajala OO. Microbial assessment and prevalence of antibiotic resistance in polluted Oluwa River, Nigeria. Egyptian Journal of Aquatic Reseaarch. 2014;**40**:291-299

[34] Hayward C, Ross KE, Brown MH, Whiley H. Water as a source of antimicrobial resistance and healthcare-associated infections. Pathogens. 2020;**9**:1-21

[35] Davies J, Davies D. Origins and evolution of antibiotic resistance. Microbiology and Molecular Biology Reviews. 2010;**74**:417-433

[36] Bai X, Ma X, Xu F, Li J, Zhang H, Xiao X. The drinking water treatment process as a potential source of affecting the bacterial antibiotic resistance. Science Total Environment. 2015;**533**:24-31

[37] Cizmas L, Sharma VK, Gray CM, McDonald TJ. Pharmaceuticals and personal care products in waters: Occurrence, toxicity, and risk. Environmnetal Chemical Letters. 2015;**13**:381-394

[38] Evgenidou EN, Konstantinou IK, Lambropoulou DA. Occurrence and removal of transformation products of PPCPs and illicit drugs in wastewaters: A review. Science Total Environment. 2015;**505**:905-9026

[39] Łuczkiewicz A, Jankowska K, Fudala-Ksiazek S, Olańczuk-Neyman K. Antimicrobial resistance of fecal indicators in municipal wastewater treatment plant. Water Research. 2010;**44**:5089-5097

[40] Mashe T, Domman D, Tarupiwa A, Manangazira P, Phiri I, Masunda K, et al. Highly resistant cholera outbreak strain in Zimbabwe. New England Journal of Medicine. 2020;**383**:687-689

[41] Mashe T, Gudza-Mugabe M, Tarupiwa A, Munemo E, Mtapuri-Zinyowera S, Smouse SL, et al. Laboratory characterisation of Salmonella enterica serotype Typhi isolates from Zimbabwe, 2009-2017. BMC Infectious Diseases. 2019;**19**:1-9

[42] Davis WW, Chonzi P, Masunda KP, Shields LM, Mukeredzi I, Manangazira P, et al. Notes from the Field: Typhoid Fever Outbreak — Harare, Zimbabwe, October 2016–March 2017. MMWR. Morbidity and Mortality Weekly Report. 2018;**68**

[43] Spengler P, Körner W, Metzger JW. Substances with estrogenic activity in effluents of sewage treatment plants in southwestern Germany. 1. Chemical analysis. Environmental Toxicology and Chemistry. 2001;**20**:2133-2141

[44] Wang L, Ying GG, Zhao JL, Liu S, Yang B, Zhou LJ, et al. Assessing estrogenic activity in surface water and sediment of the Liao River system in northeast China using combined chemical and biological tools. Environmental Pollution. 2011;**159**:148-156

[45] Lebreton LCM, Van Der Zwet J, Damsteeg JW, Slat B, Andrady A, Reisser J. River plastic emissions to the world's oceans. Nat Communications. 2017;**8**:1-10

[46] Singh S, Trushna T, Kalyanasundaram M, Tamhankar AJ, Diwan V. Microplastics in drinking water: A macro issue. Water Supply. 2022;**22**:5650-5674

[47] Jambeck JR, Geyer R, Wilcox C, Siegler TR, Perryman M, Andrady A, et al. Plastic waste inputs from land into the ocean. Science. 2015;**347**:768-771

[48] Hidalgo-Ruz V, Gutow L, Thompson RC, Thiel M. Microplastics in the marine environment: A review of the methods used for identification and quantification. Environmental Science & Technology. 2012;**46**:3060-3075

[49] Van Cauwenberghe L, Devriese L, Galgani F, Robbens J, Janssen CR. Microplastics in sediments: A review of techniques, occurrence and effects. Marine Environmental Research. 2015;**111**:5-17

[50] Pitt JA, Kozal JS, Jayasundara N, Massarsky A, Trevisan R, Geitner N, et al. Uptake, tissue distribution, and toxicity of polystyrene nanoparticles in developing zebrafish (Danio rerio). Aquatic Toxicology. 2018;**194**:185-194

[51] Anderson PJ, Warrack S, Langen V, Challis JK, Hanson ML, Rennie MD. Microplastic contamination in Lake Winnipeg, Canada. Environmental Pollution. 2017;**225**:223-231

[52] Horton AA, Walton A, Spurgeon DJ, Lahive E, Svendsen C. Microplastics in freshwater and terrestrial environments: Evaluating the current understanding to identify the knowledge gaps and future research priorities. Science Total Environment. 2017;**586**:127-141

[53] Zeenat EA, Bukhari DA, Shamim S, Rehman A. Plastics degradation by microbes: A sustainable approach. Journal of King Saudi University. 2021;**33**:101538

[54] Estahbanati S, Fahrenfeld NL. Influence of wastewater treatment plant discharges on microplastic concentrations in surface water. Chemosphere. 2016;**162**:277-284

[55] Conkle JL, Báez Del Valle CD, Turner JW. Are we underestimating microplastic contamination in aquatic environments? Environmental Management. 2018;**61**:1-8

[56] Peters CA, Bratton SP. Urbanization is a major influence on microplastic ingestion by sunfish in the Brazos River Basin, Central Texas, USA. Environmental Pollution. 2016;**210**:380-387

[57] Saad D, Chauke P, Cukrowska E, Richards H, Nikiema J, Chimuka L, et al. First biomonitoring of microplastic pollution in the Vaal river using Carp fish (Cyprinus carpio) as a bio-indicator. Science Total Environment. 2022;**836**:155623

[58] Ditlhakanyane BC, Ultra VU, Mokgosi MS. Microplastic load in the surface water and Tilapia sparrmanii (Smith, 1840) of the river systems of Okavango Delta,Botswana. Environment Monitoring Assess. 2022;**194**:1-7

[59] Nwonumara GN, Okoro PO, Okogwu OI. Assessment of the incidence of microplastics at Ndibe, Cross River, Nigeria. Zoology. 2021;**19**:46-51

[60] Migwi FK, Ogunah JA, Kiratu JM. Occurrence and spatial distribution of microplastics in the surface Waters of Lake Naivasha, Kenya.

Environmental Toxicological Chemistry. 2020;**39**:765-774

[61] Bittner M, Štern A, Smutná M, Hilscherová K, Žegura B. Cytotoxic and genotoxic effects of cyanobacterial and algal extracts–microcystin and retinoic acid content. Toxins. 2021;**13**:107

[62] Florczyk M, Łakomiak A, Woźny M, Brzuzan P. Neurotoxicity of cyanobacterial toxins. Environmental Biotechnology. 2014;**10**:26-43

[63] Picardo M, Núñez O, Farré M. Suspect and target screening of natural toxins in the Ter River catchment area in NE Spain and prioritisation by their toxicity. Toxins. 2020;**12**:752

[64] Bouaïcha N, Miles CO, Beach DG, Labidi Z, Djabri A, Benayache NY, et al. Structural diversity, characterization and toxicology of microcystins. Toxins (Basel). 2019;**11**:714

[65] Carmichael WW, Azevedo SMFO, An JS, Molica RJR, Jochimsen EM, Lau S, et al. Human fatalities form cyanobacteria: Chemical and biological evidence for cyanotoxins. Environmental Health Perspectives. 2001;**109**:663-668

[66] Paerl HW, Huisman J. Climate change: A catalyst for global expansion of harmful cyanobacterial blooms. Environmental Microbiology Reports. 2009;**1**:27-37

[67] WHO. Guidelines for drinking-water quality. Second edition, Addendum to Volume 2, Health Criteria and Other Supporting Information. Geneva: World Health Organization; 1998

[68] DWAF (Department of Water Affairs and Forestry of the Republic of South Africa). South African water quality guidelines. 2nd ed. In: Agricultural Use: Livestock Watering. Pretoria, South Africa. 1996

[69] Johansson S, Olsson M. Investigations of Toxic Algal Blooms in a Drinking Water Reservoir of Harare. Lund, Sweden: University of Lund; 1988

[70] Codd GA, Azevedo SMFO, Bagchi SN, Burch M, Carmichael WW, Harding W, et al. "CYANONET" A Global Network for Cyanobacterial Bloom and Toxin Risk Management. Initial Situation Assessment and RecommendationsNt. Hydrol Progr VI (Unesco, Paris). Hydrology. 2005;**2005**:138

[71] Samdal IA, Ballot A, Løvberg KE, Miles CO. Multihapten approach leading to a sensitive ELISA with broad cross-reactivity to microcystins and nodularin. Environmental Science & Technology. 2014;**48**:8035-8043

[72] Wang H, Wang P, Zhao X, Zhang W, Li J, Xu C, et al. What triggered the Asian elephant's northward migration across southwestern Yunnan? Innovation. 2021;**2**:100142

[73] Rzymski P, Poniedziałek B, Karczewski J, Rzymski MP. Gastroenteritis and liver carcinogenesis induced by cyanobacterial toxins Wpływ toksyn sinicowych na indukowanie ostrych nieżytów żołądka i kancerogenezy wątroby. Gastroenterology. 2011;**18**:159-162

[74] Zilberg B. Gastroenteritis in Salisbury. European children--a five-year study. The Central African Journal of Medicine. 1966;**12**:164-168

[75] Ndebele MR, Magadza CHD. The occurrence of microcystin-LR in Lake Chivero, Zimbabwe. Lakes Reservoir Research Management. 2006;**11**:57-62

[76] Masango MG. A comparative analysis of the cytotoxicity of cyanotoxins using in vitro (cell culture) and in

vivo (mouse) assays. Lakes Reservoir Research Management. 2007;**11**:1-124

[77] Mankiewicz-Boczek J, Gągała I, Jurczak T, Urbaniak M, Negussie YZ, Zalewski M. Incidence of microcystin-producing cyanobacteria in Lake Tana, the largest waterbody in Ethiopia. African Journal of Ecology. 2014;**53**:54-63

[78] Habtemariam H, Kifle D, Leta S, Beekman W, Lürling M. Cyanotoxins in drinking water supply reservoir (Legedadi, Central Ethiopia): Implications for public health safety. SN Applied Science. 2021;**3**:1-10

[79] Tamele IJ, Vasconcelos V. Microcystin incidence in the drinking water of mozambique: Challenges for public health protection. Toxins (Basel). 2020;**12**:1-20

[80] Badmus SO, Amusa HK, Oyehan TA, Saleh TA. Environmental risks and toxicity of surfactants: Overview of analysis, assessment, and remediation techniques. Environmental Science and Pollution Research. 2021;**28**:62085-62104

[81] Mungray AK, Kumar P. Fate of linear alkylbenzene sulfonates in the environment: A review. International Biodeterioration and Biodegradation. 2009;**63**:981-987

[82] Zhu FJ, Ma WL, Xu TF, Ding Y, Zhao X, Li WL, et al. Removal characteristic of surfactants in typical industrial and domestic wastewater treatment plants in Northeast China. Ecotoxicological Environmental Safety. 2018;**153**:84-90

[83] Collivignarelli MC, Carnevale Miino M, Baldi M, Manzi S, Abbà A, Bertanza G. Removal of non-ionic and anionic surfactants from real laundry wastewater by means of a full-scale treatment system. Institution of Chemical Engineers. 2019;**132**:105-115

[84] Fei X, Li W, Zhu S, Liu L, Yang Y. Simultaneous treatment of dye wastewater and surfactant wastewater by foam separation: Experimental and mesoscopic simulation study. Separation Science Technology. 2018;**53**:1604-1610

[85] Petrie B, Barden R, Kasprzyk-Hordern B. A review on emerging contaminants in wastewaters and the environment: Current knowledge, understudied areas and recommendations for future monitoring. Water Research. 2015;**72**:3-27

[86] Hampel M, Mauffret A, Pazdro K, Andalusian JB. Anionic surfactant linear alkylbenzene sulfonates (LAS) in sediments from the Gulf of Gdansk (southern Baltic Sea, Poland) and its environmental implications. Environmental Monitoring and Assessment. 2012;**184**:6013-6023

[87] Moura AGL, Centurion VB, Okada DY, Motteran F, Delforno TP, Oliveira VM, et al. Laundry wastewater and domestic sewage pilot-scale anaerobic treatment: Microbial community resilience regarding sulfide production. Journal of Environmental Management. 2019;**251**:109495

[88] Borghi CC, Fabbri M, Fiorini M, Mancini M, Ribani PL. Magnetic removal of surfactants from wastewater using micrometric iron oxide powders. Separation Purification Technolo;logy. 2011;**83**:180

[89] Zanoletti A, Federici S, Borgese L, Bergese P, Ferroni M, Depero LE, et al. Embodied energy as key parameter for sustainable materials selection: The case of reusing coal fly ash for removing anionic surfactants. Journal of Cleaned Products. 2017;**141**:230-236

[90] Martin MH, Coughtrey PJ. Biological indicators of natural ore-bodies: Geobotanical and biogeochemical prospecting for heavy metal deposits. Biology Monitoring Heavy Metal Pollution. 1982;**1982**:34-59

[91] Gavrilescu M, Demnerová K, Aamand J, Agathos S, Fava F. Emerging pollutants in the environment: Present and future challenges in biomonitoring, ecological risks and bioremediation. New Biotechnology. 2015;**32**:147-156

[92] Patel N, Khan ZA, Shahane S, Rai D, Chauhan D, Kant C, et al. Emerging pollutants in aquatic environment: Source, effect, and challenges in biomonitoring and bioremediation- A review. Pollution. 2020;**6**:99-113

[93] López-Pedrouso M, Varela Z, Franco D, Fernández JA, Aboal JR. Can proteomics contribute to biomonitoring of aquatic pollution? A critical review. Environmental Pollution. 2020;**267**

[94] Broeg K, Westernhagen HV, Zander S, Körting W, Koehler A. The "bioeffect assessment index" (BAI): A concept for the quantification of effects of marine pollution by an integrated biomarker approach. Marine Pollution Bulletin. 2005;**50**:495-503

[95] Smit MGD, Bechmann RK, Hendriks AJ, Skadsheim A, Larsen BK, Baussant T, et al. Relating biomarkers to whole-organism effects using species sensitivity distributions: A pilot study for marine species exposed to oil. Environmental Toxicology. 2009;**28**:1104-1109

[96] Masola B, Chibi M, Kandare E, Naik YS, Zaranyika MF. Potential marker enzymes and metal-metal interactions in Helisoma duryi and Lymnaea natalensis exposed to cadmium. Ecotoxicology and Environmental Safety. 2008;**70**:79-87

[97] Siwela AH, Nyathi CB, Naik YS. A comparison of metal levels and antioxidant enzymes in freshwater snails, Lymnaea natalensis, exposed to sediment and water collected from Wright Dam and Lower Mguza Dam, Bulawayo, Zimbabwe. Ecotoxicological Environmental Safety. 2010;**73**:1728-1732

[98] Siwela AH, Nyathi CB, Naik YS. Metal accumulation and antioxidant enzyme activity in C. gariepinus, Catfish, and O. mossambicus, tilapia, collected from lower Mguza and Wright Dams, Zimbabwe. Bulletin of Environmental Contamination and Toxicology. 2009;**83**:648-651

[99] Mabika N, Barson M. Histological assessment of gill pathology in two fish species (*Clarias gariepinus* and *Oreochromis niloticus*) from the Sanyati Basin in Lake Kariba, Zimbabwe. International Journal of Development and Sustainability. 2013;**2**:1476-1486

Chapter 5

# Natural Water Reservoirs as an Example of Effective Nature-Based Solutions (NBS)

*Ewelina Widelska, Barbara Sowińska-Świerkosz and Wojciech Walczak*

## Abstract

Nature-based solutions (NBS) include actions that are inspired and/or powered by nature. The level of human intervention can vary from no or minimum intervention to the creation of the entire new ecosystems. One of the types of such solutions are natural water reservoirs (NWRs) with recreational and bathing functions, in which natural water self-purification processes are used. Mechanical, biological, and chemical self-purification processes are used to filter water in natural swimming pools. The elimination of nutrients (nutrients) and bacterial contamination takes place through the use of biological filter beds, usually planted with aquatic vegetation. Implementation of natural water reservoirs also showed a multitude of positive effects on the environment benefits including: enhancing the natural capital, promoting biodiversity, creating new habitats, mitigating water runoff, enhancing water resilience, contribution to urban heat island (UHI) mitigation, increasing air quality, and improvement of local climate.

**Keywords:** nature-based solutions, natural water reservoirs, natural systems

## 1. Introduction

Intensive urban development has an increasingly stronger impact on the non-urbanized environment. At the same time, cities and their inhabitants face a huge scale of challenges, such as: air pollution, the existence of the urban heat island (UHI), water excess or scarcity, loss of natural habitats, or social stratification. The use of Nature-based solutions is an effective means of dealing with many of these problems simultaneously [1].

The concept of Nature-based solutions (NBS) refers to solutions that are powered by nature and are implemented to manage natural systems in a way that balances benefits for both nature and society [2]. They were defined by International Union for Conservation of Nature [3] as "Actions to protect, sustainably manage and restore natural and modified ecosystems that address societal challenges effectively and adaptively, simultaneously providing human well-being and biodiversity benefit."

NBS include actions that are inspired and/or powered by nature. The level of human intervention can vary from no or minimum intervention to the creation of the entire new ecosystems [4, 5]. Therefore, as NBS can be considering establishment of protected areas or conservation zones, actions directed to controlling urban expansion, gardens and parks of different size, green roofs, and facades as well as creation of new waterbodies [6].

One type of such artificial waterbodies are natural water reservoirs (NWRs) with recreational and bathing functions, in which natural water self-purification processes are used. Mechanical, biological, and chemical self-purification processes are used to filter water in natural swimming pools. The elimination of nutrients and bacterial contamination takes place through the use of biological filter beds, usually planted with aquatic vegetation. Specified, specially selected for water parameters, mineral deposits are able to capture pollutants, pathogens as well as nutrients dissolved in water, mainly phosphates [7]. Therefore, NWR meets the most important criterion set to NBS: the conscious use of natural minerals and plants is a priority, not a supplement to conventional infrastructure.

Secondly, NBS are called as challenge-orientation action that contributes to alleviate well-defined environmental, societal, and economic problems [8]. Among them, however, IUCN [3] emphasized that NBS must effectively address societal challenges as well as result in a net gain to biodiversity and ecosystem integrity. Among most commonly mentioned challenge areas addressed by NBS are climate change adaptation and mitigation, disaster risk reduction, food and water security, human health and socioeconomic development, as well as environmental degradation and biodiversity loss [6, 9, 10]. Although the main goal of natural water reservoirs' implementation is directed to societal challenge, which can be named as "health and wellbeing," these waterbodies contribute to many others challenges. Regarding the environmental dimension, these waterbodies donate biodiversity enhancement by the introduction of native plant species associated with the aquatic environment. These plants together with water area create new aquatic and semi-aquatic habitats that attract animal species, including amphibians, reptiles, and migrating birds [11]. Next to habitat creation function, natural water reservoirs face challenges such as disaster risk reduction and water management. They fulfill function of small retention waterbodies that collect rain and storm water and prevent against rapid water runoff from the urbanized areas [11]. Regarding the climate change adaptation and mitigation challenge, natural water reservoirs have direct impacts on greenhouse gas emissions via carbon storage and sequestration in vegetation, temperature cooling, and creation of shadows that impact human thermal comfort [12].

Implementation of natural water reservoirs also showed a multitude of positive effects on the environment benefits including: enhancing the natural capital, promoting biodiversity, creating new habitats, mitigating water runoff, enhancing water resilience, contribution to UHI mitigation, increasing air quality, and improvement of local climate [13]. The benefits provided by NBS may include different spatial scales: from area under action, through the surrounding areas, to the cityscape or even regional scale IUCN [3]. NWR are well known to provide a set of distinct benefits in relation to different special scales. Among cultural services are recognized benefits such as improvement of physical and mental health, increase of recreational value and educational opportunities, as well as tangible esthetic and spiritual benefits resulting from the contact with nature and people [11].

## 2. Materials and methods

Despite its significant potential, blue and green infrastructure is still insufficiently researched in Polish conditions and, therefore, remains little used as a means of counteracting the effects of climate change and adapting our cities.

### 2.1 Study area

As the study area was selected the water system within the municipal park in Zduńska Wola (Poland) - 51° 35 '52.9 "N/18 ° 55' 58.2" E. The park is located in the very center of the city, in the valley of the Pichna River, flowing through the city in the east–west direction. It is the largest green area in the city, covering an area of about 10 ha. It was established at the turn of the twentieth century as a private garden of Zenon Anstadt, the owner of the adjacent brewery. The factory owner made the park available to residents over time.

The design concept assumed the rebuilding of municipal ponds together with the surrounding area, with particular emphasis on the optimal solution for sealing, filling, and maintaining the purity of water in the ponds. The park is characterized by varied topography, the highest point of which is in the north and slopes to the south, toward the Pichna River valley, supplying two park ponds that are the closure of the compositional axis. In terms of nature, the area has significant values due to old trees and the water system.

As part of the preparatory work for the revalorization of the park, a number of studies and analyses were carried out, including assessment of the sanitary state of the waters of the Pichna River that supplies reservoirs and flows through the park. On this basis, it turned out that degree of the river pollution (among others, as a result of the discharge of untreated water from nearby traffic routes) makes it impossible to restore the water system while further supplying the ponds with river water. Therefore, it was decided to use complex and modern technological solutions that enable the renovation of the water system to ensure a satisfactory degree of purity and transparency of the water in the ponds. The idea of rebuilding the existing ponds into scenic and recreational reservoirs was based on Western European solutions used in municipal swimming ponds with natural water purification. It consisted of, inter alia, on the appropriate shape of the reservoirs, the use of natural mineral and plant filters and proper water circulation in order to repeatedly intensify the natural processes of water self-purification. In addition, the appropriate shaping of the shoreline and the introduction of swamp vegetation ensured a natural appearance of the ponds, which blend harmoniously with the surrounding landscape, which significantly contributed to the attractiveness of the entire park.

The ponds cover the area of 10,067 $m^2$ and consist of a recreation and viewing area and a mineral and plant filter. The solution is based on the use of the sorption capacity of a substrate consisting of natural ion-exchange minerals and absorbents of nitrogen, phosphorus, and heavy metals, as well as aquatic and rush plants. Water filtration technology based on the German FLL [14] standards for swimming ponds was used to ensure the best water parameters. The water surface area in the viewing and recreation area is approx. 3952 $m^2$, and the area covered with water and rush vegetation is 2664 $m^2$. The surface of the water in the pond with the mineral-plant filter is 757 $m^2$, and the area of the mineral-plant filter planted with water and rush vegetation is 2694 $m^2$, which is a total of 53% of the area planted with plants.

## 2.2 Methodology

The design concept assumed the rebuilding of municipal ponds together with the surrounding area, with particular emphasis on the optimal solution for sealing, filling, and maintaining the purity of water in the ponds. Therefore, in 2015, it was developed detailed design documentation, taking into account a series of processes that promote the natural purification of water in ponds. In spring 2017, the project was started. Prior to profiling the bottom of the ponds, the water and marsh vegetation as well as permanent weeds were removed, and the concrete slabs that had previously protected the pond slopes were removed. The subsequent step was to remove the layer of silts and peat from the bottom of the ponds and to prism them for the use in coastal areas. Appropriate control and drainage was installed at the bottom of the ponds to monitor subcutaneous waters during sealing of the ponds. Then, the bottom of ponds was properly shaped in accordance with the designed profiling. Each 30-cm layer of soil was compacted so that the formed slopes would not erode at a later stage of use.

Surface of the bottom of the pond had to be leveled and thoroughly compacted (larger depressions were buried in layers and compacted), free of sharp objects—stones, glass, etc. Sealing of the pond consisted in spreading 400 $g/m^2$ of geotextile. For a proper sealing of the pond, EPDM foil 1.02 mm thick with a certificate of neutrality with respect to flora and fauna in the ponds was used (Photo 1). In order to maintain the water table at a constant, designed level, it was necessary to permanently stabilize the edges of the pond at the appropriate level (deviations from the level could not be more than 1 cm). The entire bottom of the tanks, including the pond shore (about 1 m wide), was lined with twice washed gravel (16–32 mm fraction) with a 20-cm-thick layer.

The key element of the entire water system conditioning the purification of water in the ponds was the implementation of the mineral-plant filter (regeneration tank). Purification of water in ponds is based primarily on the proper selection of filtration material (mineral in properly developed proportions adapted to the content of elements in the water supplying the tanks) and forcing water circulation in a closed system, equipped with pumps, overflows, mechanical-mineral filter and relevant from phyto- and rhizo-filtration point of view—swamp filter (mineral bed planted with plants).

Water quality, when using biologically active sorption filters, depends on the material used and its adsorption properties and intensity of biochemical processes. It is extremely important in the process of effective removal of organic compounds (above all, phosphorus and nitrogen compounds) as well as elements present in trace amounts (e.g. heavy metals) [15]. Therefore, the mineral substrate Biozamonit® (4–16 mm fraction) was used, which is the nitrogen, phosphorus, and heavy metals as well as parasites sorbent with the addition of FerroSorp® iron hydroxide (phosphorus absorbent). This trade name includes lime-silica rocks (bedrock), zeolite, dolomite, or limestone grits in appropriate proportions [16].

Due to the high calcium content, it is the bedrock that is one of the most effective reactive materials used to remove phosphorus from aqueous solutions [17]. Furthermore, during thermal treatment, its sorption capacity increases significantly (from 60.5 g P•kg −1 at 250°C to 119.6 g P•kg −1 at 1000°C) [18] due to the breakdown of calcium carbonate into calcium oxide and carbon dioxide. As the firing temperature increases, the reaction also increases: from 6.80 to 12.4 after roasting at 900–1000°C [18, 19].

**Figure 1.**
*Mineral-plant filter with visible plantings of purple loosestrife, bulrush, and Calamus (photo: W. Walczak).*

Water pollution can be removed using mechanical, physicochemical, and biological methods. Therefore, in parallel with the use of physicochemical methods, a biologically uncomplicated method, which is phytoremediation, was used. It is a technology for the purification of ground and surface waters, and even soil and air, which uses the natural predisposition of specific taxa of plants capable of growing and developing in ecosystems contaminated with inorganic and organic substances, as well as for their uptake, accumulation, or biodegradation [20].

Therefore, several habitat zones have been arranged in the pond. In the open water, plants with delicate, flabby stems, and fine leaves (adapting their requirements to the depth of the zone) were planted (**Figure 1**). Water lily with decorative leaves and flowers were planted in the bottom layer. There are also numerous zooplankton species in this zone, which very effectively support water filtration [21]. The banks were planted with iris, calamus, and other littoral species, as well as plants from wetland habitats—sedge, horsetail, and mint [22]. All species were selected according to the habitat requirements criterion.

## 3. Results and discussion

The use of mineral-plant filtration bed in ponds is aimed at eliminating the most adverse biogenic compounds (phosphorus and nitrogen) and maintaining optimal physical and chemical parameters of water (similar to those found in natural oligotrophic reservoirs). Artificial stimulated conditions served to limit the development of unicellular and filamentous algae, determining the natural and scenic values of ponds (each additional gram of phosphorus above the allowed norm can generate the development of filamentous algae weighing 250 kg).

In the water reservoirs given in the analyses, a natural mineral substrate Biozamonit® was used as an absorbent of phosphorus, nitrogen, and heavy metals. Composition of the above mineral deposit contains, among others, bedrock, roasted rock, zeolite, calcareous grits, ferrosorp, and dolomite grits in appropriate proportions. Biozamonit® contains, among others, additions of roasted rock, which according to tests, has a contents of CaO and $SiO_2$ of 238.6 and 550.1 g•kg$^{-1}$, respectively. Therefore, the efficiency of $PO_4$-P removal from an aqueous solution with an initial concentration of 1.84 and 2.88 mg•dm$^{-3}$ by the roasted rock is 88% and 70%, respectively [23]. It is also important to place the substrate in the appropriate parts of the filter bed, because the filtration processes in the mineral-plant bed take place through the vertical forced flow of water through the mineral bed.

It was the filtration, ion exchange, and buffer properties that determined the use of the aforementioned mixture as the best for this type of pond, successfully used in other projects of reservoirs of a scenic and recreation nature. The filter bed is also a habitat supporting the growth of nitrifying bacteria that are very important in the process of ammonia decomposition.

In order to increase sorption properties, the bed was planted with water and marsh plants with the best properties of absorbing the biogenic compounds from water. Plants that are used for the biological purification of water belong to the group of the so-called repository plants, highly effective also in removing toxins, protecting banks, reclamation, and cleaning soil or water. They capture pollution, pathogens, and phosphates dissolved in water and at the same time strive to stabilize water chemistry. Among the most commonly used plants for this purpose, there are helophytes (marsh plants), emery hydrophytes, and submersible hydrophytes that

| **Physicochemical parameter tested** | **Measurement unit** | **Tap water for filling the ponds (before treatment)** | **Pond water I season (spring 2018)** | **Pond water II season (spring 2019)** |
|---|---|---|---|---|
| pH | — | 7.3 | 7.8 | 7.9 |
| Manganese | mg•dm$^{-3}$ | 0.023 | 0.01 | 0.02 |
| Ammonia | mg NH4•dm$^{-3}$ | <0.060 | 0.04 | 0.02 |
| Nitrate | mg NO3•dm$^{-3}$ | 2.05 | 0.8 | 0.5 |
| Nitrite | mg NO2•dm$^{-3}$ | 0.01 | 0.02 | 0.02 |
| Electrical conductivity | µS•cm$^{-1}$ | 452 | 357 | 223 |
| Turbidity | NTU | 0.51 | 0.8 | 0.65 |
| Total hardness | dH | 12.65 | 7 | 6 |
| Carbonate hardness | dH | 4 | 6 | 6 |
| Color | mg•dm$^{-3}$ Pt | 5 | 8 | 8 |
| Chloride | mg•dm$^{-3}$ | <5.0 | 5.5 | 8.6 |
| Sulfide | mg SO4$^{2-}$ •dm$^{-3}$ | <20 | < 20 | < 20 |
| Soluble phosphorus | mg•dm$^{-3}$ | 0.05 | 0.006 | 0.005 |
| Total phosphorus | mg•dm$^{-3}$ | 0.25 | 0.035 | 0.025 |

**Table 1.**
*Summary of water test results.*

live completely under the water [24]. Both reservoirs were planted with aquatic and underwater plants, including those with a covering function.

As a result of the sorption properties of the mineral-plant filtration bed (regeneration tank) in the first and second year of using the ponds, the amount of dissolved phosphorus (P), one of the biogenic factors, decreased 10-fold compared to the amount contained in tap water used to fill the ponds. Thus, as shown in **Table 1**, it has come close to the European standards regarding the purity of water in bathing tanks. The amount of nitrates decreased fourfold and ammonium threefold. The carbonate hardness remains at the right level around 6 dH, and the pH of the water is optimal. The complete elimination of harmful nutrients is impossible, because they are periodically influenced by the evapotranspiration process, feces of water birds, plant pollen, or small particles of airborne pollutants. As experience from other water reservoirs with scenic and recreational functions shows, the use of effective mineral and plant beds with appropriately selected physicochemical properties contribute to the effective filtration of water in ponds.

## 4. Conclusions

Significant benefits of the construction of retention ponds are the possibility of collecting water for use in periods of drought, providing habitats for wild plants and animals in urbanized areas, and functional and compositional enhancement of public green areas. An additional benefit is the ability to purify water from contaminants from surface runoff through sedimentation and phytoremediation.

In order to increase the usability of the pond, especially in urbanized areas (**Figure 2**), it is worth designing it in such a way as to enable recreational use by various age groups of users as well as to provide the opportunity to conduct environmental education. The location of the reservoir should contribute to the improvement of the continuity of naturally functioning areas and thus favor the ecological role of this element, significantly increasing biodiversity in the urban environment.

**Figure 2.**
*Pond before renovation (photo: Milecka M.).*

**Figure 3.**
*Recreational pond after 2 years of operation (photo: Walczak W.).*

The observations to date indicate a very high efficiency of natural self-purification of water in bathing reservoirs [7]. This is possible thanks to the enormous vital activity of the biocenosis of natural swamp systems—from microorganisms to plants and algae, which are in dynamic balance with each other. They trigger a number of physical, chemical, and biochemical processes such as sedimentation, adsorption, oxidation, and the exchange of volatile substances between the atmosphere and water (release of gaseous metabolic products into the atmosphere), which, by interacting with each other, maintain the biological balance in the pond (**Figure 3**). The reduction of coliforms or enterococci on properly designed filters is particularly effective, so much so that the standards required by sanitary regulations are usually achieved with reserve [25]. On the other hand, one of the most difficult technical challenges posed by such solutions is to maintain the phosphate concentration in the top-up and bathing water at the level characteristic of oligotrophic waters [7]. In terms of efficiency, the analyzed joints are cheaper to maintain and implement than conventional solutions.

**Author details**

Ewelina Widelska*, Barbara Sowińska-Świerkosz and Wojciech Walczak
University of Life Science in Lublin, Poland

*Address all correspondence to: e.wid@wp.pl

## References

[1] Iwaszuk E, Rudik G, Duin L, Mederake L, Davis McK, Naumann S, (Ecologic Institute); Wagner I, (FPP Enviro), Blue-Green Infrastructure for Climate Change Mitigation - Technical Catalog. Copyright by Ecologic Institute & Fundacja Sendzimira Berlin – Kraków; 2019. 104 p

[2] Science for Environment Policy. The Solution Is in Nature. Future Brief 24. Bristol: Brief produced for the European Commission DG Environment, Science Communication Unit, UWE Bristol; 2021

[3] IUCN. Global Standard for Nature-Based Solutions. A User-Friendly Framework for the Verification, Design and Scaling up of NbS. First ed. Gland, Switzerland: IUCN; 2020

[4] Eggermont H, Balian E, Azevedo JMN, Beumer V, Brodin T, Claudet J, et al. Nature-based solutions: New influence for environmental management and research in Europe. In: GAIA - Ecological Perspectives for Science and Society. Vol. 24. oekom verlag; 2015. pp. 243-248

[5] Somarakis G, Stavros S, Chrysoulakis N, editors. Think Nature Nature-Based Solutions Handbook, Think Nature Project Funded by the EU Horizon 2020 Research and Innovation Programme under Grant Agreement No 730338. European Union; 2020

[6] Dumitru A, Wendling L, editors. Evaluating the Impact of Nature-Based Solutions: A Handbook for Practitioners. D-G R&I. Luxembourg: European Commission; 2021. DOI: 10.2777/244577

[7] Widelska E, Walczak W. Restoration of ponds in the municipal park in Zduńska Wola, Poland. Journal of Water and Land Development. 2020; **No. 44**(I–III):151-157. DOI: 10.24425/jwld.2019.127056

[8] Sowińska-Świerkosz B, Garcia J. A new evaluation framework for nature-based solutions (NBS) projects based on the application of performance questions and the indicators approach. Science of the Total Environment. 2021;**787**:147615. DOI: 10.1016/j.scitotenv.2021.147615

[9] European Commission. Towards an EU Research and Innovation Policy Agenda for Nature-Based Solutions & Re-Naturing Cities. Brussels: European Commission; 2015

[10] United Nations Environment Programme. Adaptation Gap Report 2020. Nairobi: United Nations Environment Programme; 2021

[11] Milecka M, Widelska E. Problem zanieczyszczenia rzeki Pichny w kontekście rewaloryzacji parku miejskiego w Zduńskiej Woli. Acta Scientiarum Polonorum. Formatio Circumiectus. 2018;**17**(1) Publishing House of the University of Agriculture in Krakow:59-73. DOI: 10.15576/ASP.FC/2018.17.1.67

[12] Falkenmark M. Water resilience and human life support - global outlook for the next half century. International Journal of Water Resources Development. 2020;**36**(2-3):377-396. DOI: 10.1080/07900627.2019.1693983

[13] Votruba L, Broža V. Water Management in Reservoirs, Developments in Water Science. Vol. 331989. Elsevier

[14] Mahabadi M. Richtlinien für Planung, Bau und Instandhaltung von Fassadenbegrünungen,

Fassadenbegrünungsrichtlinien (norm FLL). 2017

[15] Pruss A, Maciołek A, Lasocka-Gomuła I. Wpływ aktywności biologicznej złóż węglowych na skuteczność usuwania związków organicznych z wody [effect of the biological activity of carbon filter beds on organic matter removal from water]. Ochrona Środowiska. Vol. 31. Nr 4. 2009. pp. 31-34

[16] Walczak W. Szczegółowe specyfikacje techniczne wykonania i odbioru robót związanych z inwestycją pod nazwą, Projekt rewaloryzacji parku miejskiego w Zduńskiej Woli Stawy parkowe [Detailed technical specifications for the implementation and acceptance of works related to the investment under the name "Project of revalorization of the city park in Zduńska Wola" – Park ponds]. Zduńska Wola, Urząd Miasta; 2015. p. 45

[17] Bus A, Karczmarczyk A. Charakterystyka skały wapienno-krzemionkowej opoki w aspekcie jej wykorzystania jako materiału reaktywnego do usuwania fosforu z wód i ścieków [Properties of lime-siliceous rock opoka as reactive material to remove phosphorous from water and wastewater]. Infrastruktura i Ekologia Terenów Wiejskich. Nr. 2014;**II**(1):227-238

[18] Brogowski Z, Renman G. Characterization of opoka as a basis for its use in wastewater treatment. Polish Journal of Environmental Studies. 2004;**13**(1):15-20

[19] Cucarella V, Zaleski T, Mazurek R. Phosphorus sorption capacity of different types of opoka. Annals of Warsaw University of Life Sciences – SGGW. Land Reclamation. Nr 38. 2007. pp. 11-18

[20] Koźmińska A, Hanus-Fajerska E, Muszyńska E. Możliwość oczyszczania środowisk wodnych metodą ryzofiltracji [Possibilities of water purification using the rhizofiltration method]. Woda-Środowisko-Obszary Wiejskie. 2014; **T. 14. Z. 3**(47):89-98

[21] Spieker J, von Teichen G. Die wesentlichen Einflussfaktoren auf die Wasserqualität [aquatic ecology of ponds. The main factors influencing water quality]. Schwimm- und Gartenteiche. GaLaBau. Nr. 2008;**3**(08):56-59

[22] Milecka M, Widelska E, Walczak W, Gąsiorowski M, Jendryka J, Antończak A. Opracowanie dokumentacji projektowej na zagospodarowanie wód deszczowych oraz renowację zbiorników wodnych na terenie Parku Miejskiego w Zduńskiej Woli [Development of design documentation for rainwater management and renovation of water reservoirs in the City Park in Zduńska Wola]. Urząd Miasta, Zduńska Wola; 2015

[23] Karczmarczyk A, Bus A. Usuwanie Fosforu z małych cieków i zbiorników Wodnych w Obszarach użytkowanych Rolniczo [Phosphorus Removal from Small Watercourses and Water Reservoirs in Agricultural Areas]. Konferencja Naukowo-Techniczna, Innowacyjne Technologie w inżynierii i kształtowaniu środowiska [Science and Technology Conference "Innovative Technologies in Engineering and Shaping the Environment"]. Lublin; 2013

[24] Gąsiorowski M. Publiczne stawy kąpielowe [Public swimming ponds] [online]. Zieleń Miejska. Dodatek specjalny. Nr 2013-9. 2013. Available from: http://eczytelnia.abrys.pl/dodatek-specjalny/2013-9-701/zeszytspecjalny-8101/

publiczne-stawy-kapielowe-16606
[Accessed: June 15, 2019]

[25] Naidoo S, Olaniran A, O. Treated wastewater effluent as a source of microbial pollution of surface water resources. International Journal of Environmental Research and Public Health. 2014;**11**(1):249-270. DOI: 10.3390/ijerph110100249

Chapter 6

# Subterranean Waters in Riviera Maya of the Yucatan Peninsula: Vulnerability and the Importance of Monitoring

*Luis M. Mejía-Ortíz, Alejandro L. Collantes-Chávez-Costa, Cruz López-Contreras and Oscar Frausto-Martínez*

## Abstract

The Mexican Caribbean coast has great scenic beauty both on the surface and underwater, which is why it has been a developing area for tourism since the 1970s, establishing sites such as Cancun and Playa del Carmen and empowering others such as Cozumel and Tulum. Their biological richness is enormous, especially in the Mesoamerican Reef of which they are a part. However, this richness and scenic beauty are not possible without the ecological assemblages that exist within these regions' adjacent ecosystems, mainly the surrounding seasonally dry tropical forest and the coastal wetlands that, together with the oceanographic characteristics of the Caribbean Sea, potentiate it, turning the region into the most visited in Latin America. To this end, groundwater plays a very important role in the assemblages of biotic and abiotic elements that are shared with the Caribbean Sea; thus, its constant monitoring allows us to identify how the changes that occur in the tropical forest are producing various changes in the composition and abundance of coastal reef elements. Here, we present results of our study of groundwater conditions (temp, pH, oxygen dissolved, and salinity) in nineteen cenotes and underground rivers of the Riviera Maya and six cenotes of Cozumel. We also profiled the predominant vegetation on the surface of this region, which is a seasonally dry tropical forest, to understand the components and functioning of these subterranean ecosystems to assess their vulnerability and identify their threats from human development (population growth, tourism development, mobility capacity). These threats not only affect the cave and coastal organisms but also the tropical karstic landscapes that are characteristic of these systems.

**Keywords:** underground, aquatic monitoring, ecological links, vulnerability

## 1. Introduction

The groundwaters in many places around the world are sources of freshwater for human use [1]. In Mexico, the Yucatan Peninsula is the largest freshwater source in the region. Its soil is mainly karstic, allowing a high porosity and filtration of rainwater to the underground [2, 3]. Due to this reservoir, the tropical forest is tall and exuberant because the soil layer is thin enough to allow tree roots to reach the aquifer. The extreme hot conditions cause the forest to be dry or semi-dry [4, 5]. Over many years, these conditions have produced thousands of conduits and entrances [6, 7] that reach to coastal areas as subterranean rivers. The richness of the Caribbean Sea establishes important conditions for the proliferation of coral reefs [8]. The terrestrial ecosystems (tropical forest, mangrove, and dune) are connected by underground waters that establish links among the adjacent terrestrial and marine ecosystems (coastal lagoons, sea grass, and coral reefs). However, this role that is important, has been poorly valuation, and the groundwaters has a unique characteristics that produce a special ecosystems [9]. The state of Quintana Roo and its coast have natural attractions with unique scenic beauty, such as Caribbean beaches, reefs, lagoons, and archeological sites, among others. Tourism is based on sun and sand, but there are other tourist attractions that are highly visited, such as nature reserves and cultural areas [10]. This set of attractions motivated the creation of Cancún in 1970, which became the largest and most important tourist destination in Mexico and the Caribbean [11]. Natural attractions also led to the development of the Riviera Maya in the 1990s, which is equally as recognized as Cancun [12]. Together, these locations receive around 12 million visitors per year [13]. Initially, tourists visited for the coastal and marine landscape, but in recent years visitors have become more interested in the cenotes and underground activities. All these, therefore that another human activity such as mobility, pork farms, soil extractions and growing of the cities are the development project that in the next years will produce environmental impacts in special to water [14]. For these reasons, it is important to monitor the water to understand how the ecosystems work.

## 2. Material and methods

From 2016 to 2020, we profiled 19 cenotes in Riviera Maya and six in Cozumel (**Figure 1**) using the Hydrolab Data Sonde 5 (Hydrolab DS5) to identify the different layers and variations. We recorded dissolved oxygen (±0.01 mg/l), pH (±001 pH), salinity (±0.01%), and water temperature (±0.01°C). The Hydrolab DS5 was programed to record every 30 seconds and the divers introduce in the cave during one and half hours. At the same time, recording of faunal composition was conducted to identify crustaceans such as isopods, different species of shrimp, amphipods, and remipedes [15–17]. Divers also recorded tree roots into the aquifer using taxonomic guides [18, 19] and identified the systems with clear water discharges to the Caribbean Sea. We also examined tourism and tourist activities in the Riviera Maya and Cozumel and corroborated this data with vegetation and soil use in the surrounding the cenotes as well as with data reported in the literature.

## 3. Results

### 3.1 Environmental conditions

After checking the profiles of twenty-four underground systems, it was possible to recognize three main ecosystems. The first is a freshwater ecosystem that is not stratified and that has similar conditions of oxygen, temperature, and pH in superficial

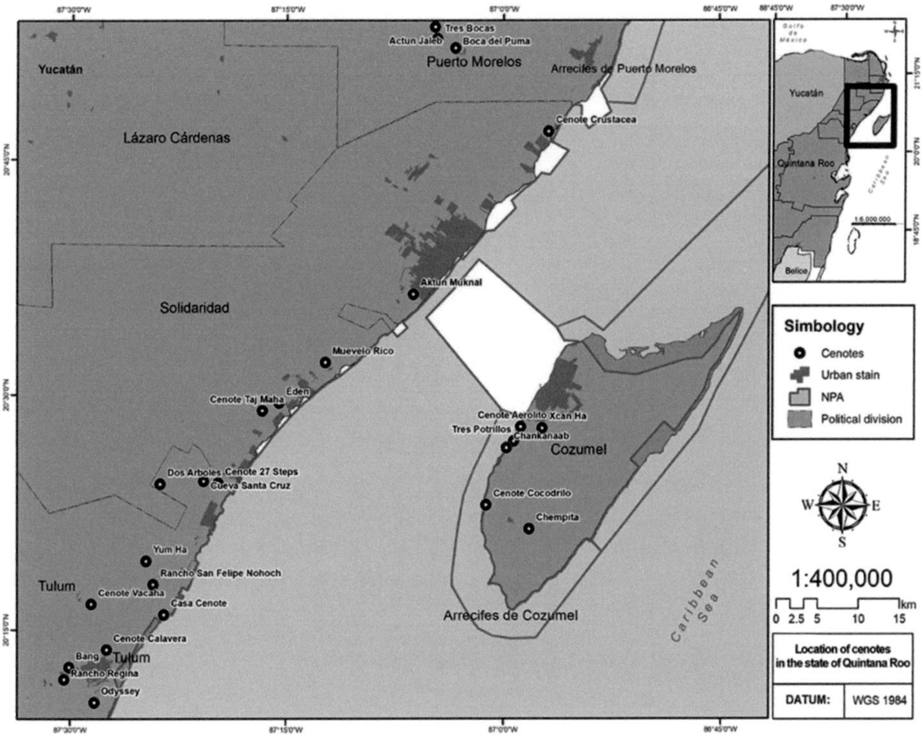


**Figure 1.**
*Location of the different cenotes in the Riviera Maya and Cozumel.*

and bottom areas. We recorded 10 sites in freshwater ecosystems (**Table 1**), some of which are shallow ponds and some of which are 15 meters deep.

In these profiles, the water column is not stratified accordingly with the salinity content. The temperature and oxygen dissolved have a negative relationship with the depth because both variables decrease as depth increases. However, pH values are more independent and showed variations, possibly due to the biochemical karst process, as has been reported for other systems [20] (**Figure 2**).

The second type of ecosystem is a brackish/marine water ecosystem. In these ecosystems the marine entrance has a direct connection with the sea. As such, some marine animals such as echinoderms and some species of fish live in caves. In these systems the mixing zone between freshwater and marine water is all time and only in some season is possible identify stratification in the water column. We recorded two systems in this study (**Figure 3** and **Table 2**).

The third type of ecosystem is an anchialine pool, which is a subterranean estuary containing freshwater and marine water separated by a mixing zone called the halocline. Anchialine pools can be found at different depths according to their distance from the sea. In this study, we recorded thirteen of these systems in the Riviera Maya and Cozumel. In these ecosystems, the stratified water column is due to the salinity changes that produce similar conditions of temperature, pH, and oxygen dissolved. It is interesting that in these environments the most rich species were found just past the halocline in the marine layer.

In anchialine ecosystems the temperature and oxygen dissolved decrease as depth increases, and both have a direct relationship with the halocline at different depths.

| Systems | Salinity | Environmental conditions | Organisms recorded |
|---|---|---|---|
| Boca del Puma | Freshwater | The water profile is not stratified with similar values of temperature and pH; decreasing a little for oxygen dissolved | *Typhlatya mitchelli;* Creaseriella anops; *Creaseria morleyi* Antromysis cenotensis |
| Tres Bocas | Freshwater | | *Typhlatya mitchelli; C. anops; Creaseria morleyi A. cenotensis* |
| Actun Jaleb | Freshwater | | *Typhlatya mitchelli; C. anops; Creaseria morleyi A. cenotensis* |
| Muevelo Rico | Freshwater | | *Typhlatya mitchelli; C. anops; Creaseria morleyi A. cenotensis* |
| Aktun Muknal | Freshwater | | *Typhlatya mitchelli; C. anops; Creaseria morleyi A. cenotensis* |
| Vacaha | Freshwater | | *Typhlatya mitchelli; C. anops; Creaseria morleyi A. cenotensis Stygomysis holthuisi* |
| Cueva Santa Cruz | Freshwater | | *Typhlatya mitchelli; C. anops; Creaseria morleyi A. cenotensis* |
| Dos Arboles | Freshwater | | *Typhlatya mitchelli; C. anops; Creaseria morleyi A. cenotensis* |
| San Felipe Nohoch | Freshwater | | *Typhlatya mitchelli; C. anops; Creaseria morleyi* |
| Yum Ha | Freshwater | | *Typhlatya mitchelli; C. anops; Creaseria morleyi A. cenotensis* |

**Table 1.**
*Record of freshwater subterranean ecosystems with environmental conditions and fauna registered.*

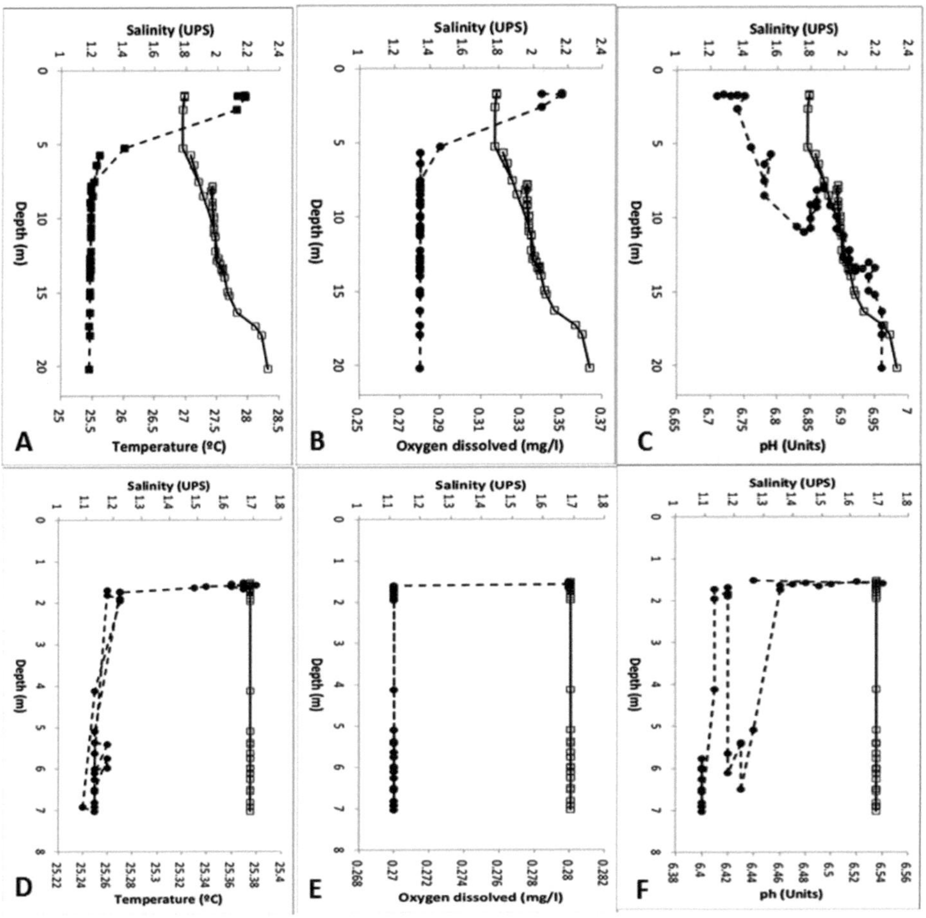


**Figure 2.**
*Examples of profiles according to the depth of freshwater ecosystems. A, B, and C graphs are for the Santa Cruz system, and D, E, and F graphs are for the Vaca ha system. Salinity is represented by continuum lines with □, and temperature, oxygen dissolved, and pH are represented by dashed lines with •.*

However, in the case of Cenote Calavera, the temperature increases at same time that the marine layer is present. In contrast, the pH values in all cenotes are higher in comparison with the superficial layer due to the alkalinity value of the marine layer (**Figure 4**).

These three main ecosystems exist in different places according with several factors such as surrounding vegetation but mainly with the depth when the subterranean branch cave reaches, but also is relationship with the distance from the coast that the cenote (entrance) is located because the incorporation of the marine layer is most important in the those systems close to the coastal area but far away to the coastal line the freshwater lenses is more wider whilst that the marine layer is found to deeper area (**Figure 5**).

### 3.2 Surrounding vegetation to the aquifer

The semi-evergreen seasonal forest is the dominant tropical forest in the study area, and in the state of Quintana Roo [21, 22]. In Mexico, this kind of seasonal

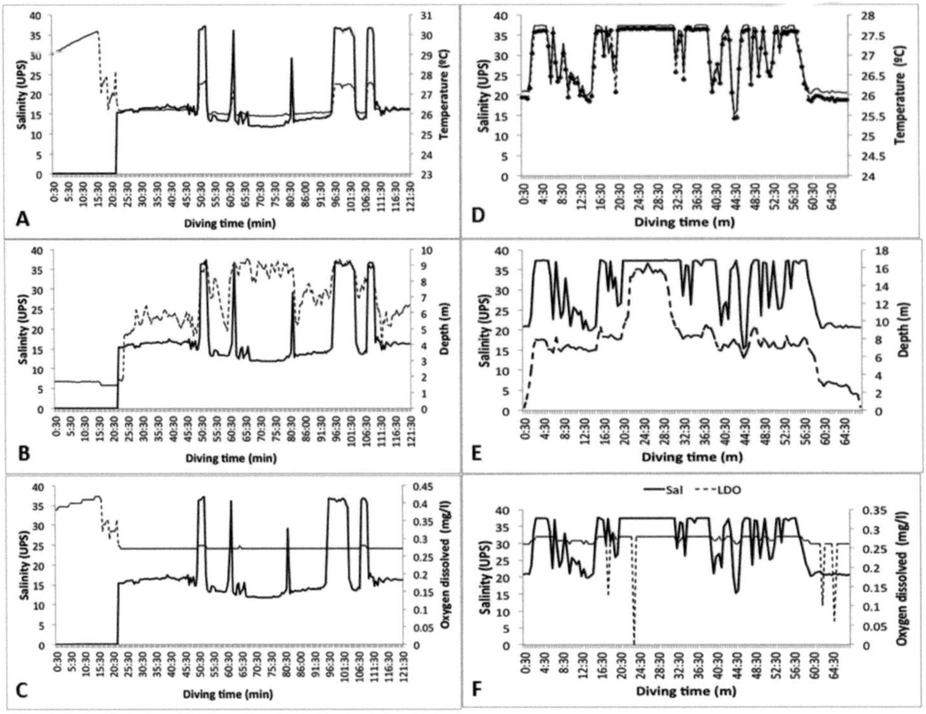


**Figure 3.**
*Examples of profiles of brackish/marine ecosystems according to diving time. A, B, and C graphs are for casa cenote from the Actun ha system, and D, E, and F graphs are for cenote Aerolito in Cozumel. Salinity is represented by continuum lines, and temperature, oxygen dissolved, and pH are represented by dashed lines with •.*

| Systems | Salinity | Environmental conditions | Organisms recorded |
|---|---|---|---|
| Casa Cenote | Brackish/Marine water | The water profile is not stratified; just the mixing zone; and the temperature, oxygen dissolved, and pH showed values similar in the superficial and deeper layers | *Typhlatya pearsei; Metacirolana mayana* |
| Aerolito | Brackish/Marine water | | *Procaris mexicana; Metacirolana mayana; Xibalbanus cozumelensis;* Copidaster cavernicola; *Ophionereis conmutabilis* |

**Table 2.**
*Record of brackish/marine water subterranean ecosystems with environmental conditions and fauna registered.*

tropical forest is known as a tropical evergreen forest [23] or medium semi-evergreen forest [24]. This vegetation develops in the *Aw* climate [25], characterized by a short but well-marked dry season, which in the area extends from November to May, during which short, sporadic, and infrequent rains may occur (191 mm in average), in addition to a rainy season the rest of the year, with rainfall around 1000 mm and an average annual temperature between 20 and 25°C. A semi-evergreen seasonal tropical forest grows on mainly flat terrain with shallow soils (20 or 35 cm deep), generally of the Rendzina type, and with outcrops of limestone [4].

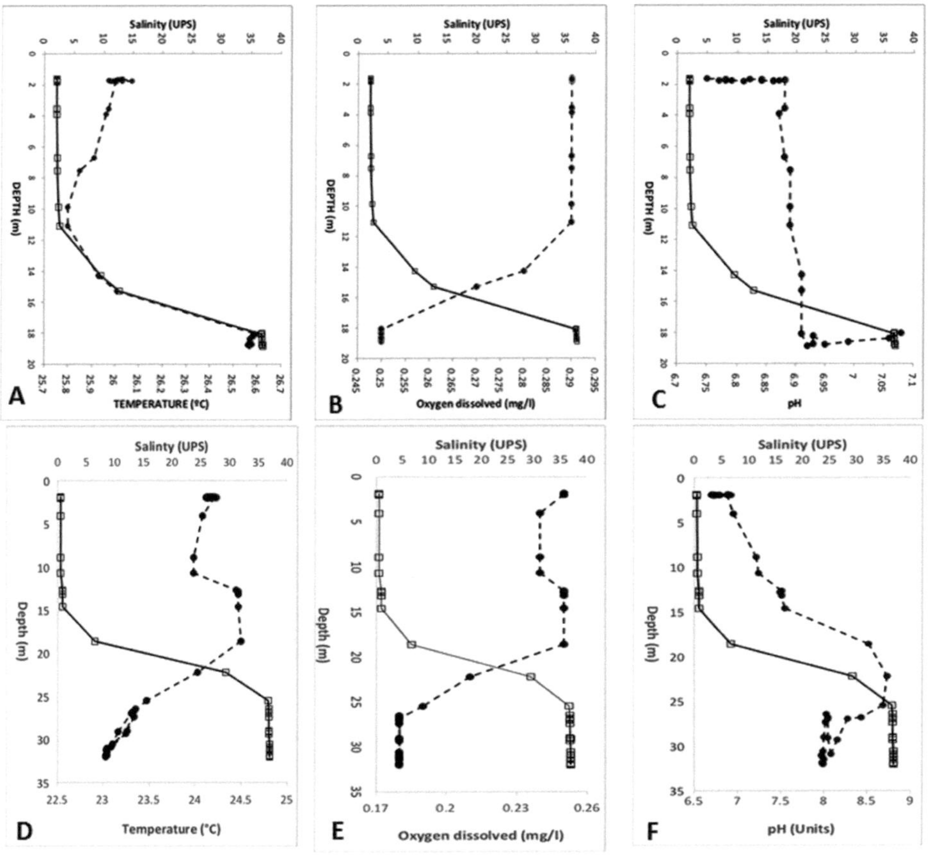


**Figure 4.**
*Examples of profiles according to the depth of the water column of anchialine ecosystems. A, B, and C graphs are for cenote Calavera, and D, E, and F graphs are for cenote Chempita in Cozumel. Salinity is represented by continuum lines with □, and temperature, oxygen dissolved, and pH are represented by dashed lines with •.*

Although a large area of this tropical seasonal forest is currently secondary, relics of the original forest can still be found [26]. Several tree strata define its vertical structure with the abundant presence of climbers and epiphytes and a well-developed shrub layer as well as an herbaceous layer composed of seedlings from the species at the upper strata [18]. The characteristic tree species are *Bursera simaruba* (L.) Sarg., *Metopium rownie* (Jacq.) Urb., *Vitex gaumeri* Greenm., *Brosimum alicastrum* Sw., *Manilkara zapota* (L.) P. Royen, and *Psidium sartorianum* (O.Berg) Nied. Among the most frequent shrubs are *Acacia collinsi* Saff., *Bauhinia jenningsii* P. Wilson, and *Eugenia acapulcensis* Steud. To name a few. In the epiphytic stratum, there are species such as *Brassavola nodosa* (L.) Lindl. And *Selenicereus testudo* (Karw. Ex Zucc.) Buxb.

The height of trees in the semi-evergreen seasonal tropical forest ranges from 18 to 25 m [24]. Trees of up to 35 m can also be found [4]; these are associated with better environmental conditions [18] such as areas with greater soil depth and karst water bodies (cenotes). The vegetation associated with these cenotes has the same floristic composition of the semi-evergreen seasonal tropical forest, but the tree stratum shows to be characteristically evergreen with a greater abundance of epiphytes (**Figure 6**) [4, 19].

**Figure 5.**
*Schematic underwater profile where it is possible to see the relationship (halocline) between freshwater (light blue) and marine (dark blue) layers, according to depth and distance from the sea. The tree roots that reaching the aquifer to maintained the semi-evergreen state of tropical forest and mangrove vegetation.*

## 4. Discussion

### 4.1 Tourist activity in the cenotes and its impacts

The great diversity and scenic beauty of the Riviera Maya, both on the surface and underwater, is related to the different types of ecosystems and large amounts of natural resources. This has facilitated the diversification of tourist activity, which now also includes exploring underground rivers, cenotes, and caves [27].

Visits to cenotes have increased significantly. The arrival of sargassum on the coast of Quintana Roo has motivated visitors to move to these spaces [28], since tourism service providers have recognized them as a substitute for beaches.

The cenotes have great landscape and cultural importance. Their underwater landscapes are geologically sculptural, which, together with their link to the general Mayan culture, creates a synergy in the visit. The cultural importance of cenotes dates

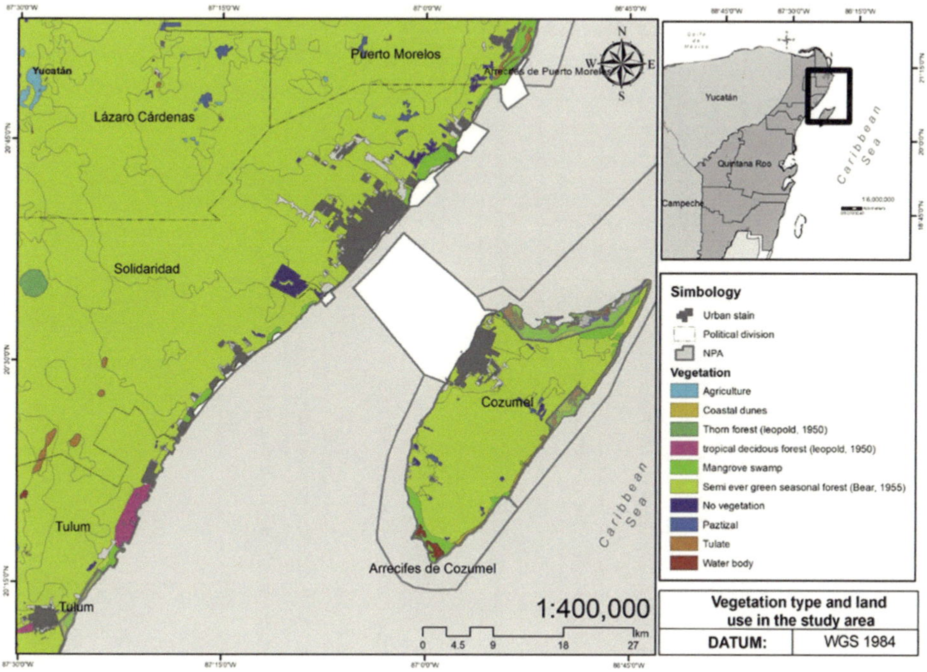


**Figure 6.**
*Vegetation type and land use for the study area.*

to pre-Hispanic times, and rituals and offerings learned from their Mayan ancestors are still performed these days to request rain, care for crops, and to ask for permission to enter the cenote [29]. Cenotes contain unique fauna and flora, formations such as stalagmites and stalactites, and ritual elements of the Mayan civilization. In addition, the areas around the cenotes are usually surrounded by jungle, mangrove, or palm and with different species of fauna that give each site an evocative power. This is why it is in these areas that observation and interpretation walks of the Mayan culture are offered.

The cenotes not only have unique esthetic natural characteristics that entice people to visit them, but they are also repositories of unique species of flora and fauna, which is where the biological and ecological importance of these systems lies [30]. Although cenotes are highly fragile ecosystems, there are currently no regulations for their use and management. In addition, since they are not isolated systems but rather they connect with groundwater that sometimes connects with the sea [31], their inappropriate use can cause unquantifiable impacts on the rest of the ecosystems with which they interact [28].

The excess of tourists in the cenotes is putting the aquifer and its aquatic biota at risk. Visitors introduce physical, chemical, and biological agents into the cenotes, such as garbage, sunscreens, repellents, creams, and fecal matter. These contaminants can alter the system, causing changes in temperature, erosion, increase in fecal coliforms, and excess nutrients that can generate irreversible impacts [27, 28]. Several of the cenotes studied experience high tourist activity (**Table 1**) because they are being offered as recreational spaces for swimming, snorkeling, diving, and hiking.

| Systems | Salinity | Environmental conditions | Organisms recorded |
|---|---|---|---|
| Cocodrilo | Anchialine; Halocline at 8 m | The water profile is stratified with lower values in deeper area (marine layer) for temperature; oxygen dissolved and pH increase to alkaline values according with the deeper area | *Tulumella sp.* |
| Chankanaab | Anchialine; Halocline at 7 m | | *Xibalbanus cozumelensis; Barbouria cubensis* Macrobrachium carcinus; Macrobrachium faustinum; |
| Tres Potrillos | Anchilaline; Halocline at 12 m | | *Agostocaris zabaletai, Procaris mexicana, Typhlatya sp., Barbouria cubensis; Calliasmata nohochi; Metacirolana mayana* |
| Taj Maha | Anchialine; Halocline at 14 m | | *Typhlatya pearsei; Xibalbanus tulumensis;* Antromysis cenotensis |
| 27 Steps | Anchialine; Halocline at 15 m | | *Xibalbanus tulumensis; Barbouria cubensis; Typhlatya pearsei; Typhlatya mitchelli; A. cenotensis* |
| Crustacea | Anchialine; Halocline at 15 m | | *Typhlatya pearsei; Xibabalbanus tulumensis; Xibalbanus fuchscockburni; Metacirolana mayana; A. cenotensis* |
| Odyssey | Anchialine; Halocline at 18 m | | *Typhlatya pearsei; Typhlatya mitchelli; A. cenotensis* |
| Bang | Anchialine; Halocline at 16 m | | *Typhlatya pearsei; Typhlatya mitchelli; A. cenotensis* |
| Calavera | Anchialine; Halocline at 18 m | | *Typhlatya pearsei; Xibalbanus tulumensis; Typhlatya mitchelli; Mayaweckelia sp.* |
| Eden | Anchialine; Halocline at 18 m | | *Typhlatya pearsei; Typhlatya mitchelli* |
| Regina | Anchialine; Halocline at 20 m | | *Xibalbanus tulumensis; Typhlatya mitchelli; Typhlatya pearsei* |
| Chempita | Anchialine: Halocline at 24 m | | *Anchialocaris paulini, Agostocaris zabaletai, Xibalbanus sp; Metacirolana mayana; Mayaweckelia sp. Tulumella sp.* |
| Xcan Ha | Anchialine; Halocline at 24 m | | *Agostocaris bozanici;* |

**Table 3.**
*Record of brackish/marine water subterranean ecosystems with environmental conditions and fauna registered.*

The three ecosystem types discussed have a strong relationship with the dynamics of tropical dry forest function, especially in those that are far away from the coast. However, these relationships also exist in coastal areas where the vegetation type is characterized by mangrove and dune. How as been reported to terrestrial cave environments they are not isolated and in these cases the cave systems fully of water have interesting relationship with the out ecosystems surrounding [32]. The freshwater from the aquifer in Quintana Roo is one of the best-conserved reservoirs in the country, but at same time it faces several threats. Exponential population growth has occurred in Playa del Carmen, Tulum, Puerto Morelos, Akumal, Cozumel, and Puerto Aventuras, thus increasing demand for water. Tourism activity is changing from sun and beach activities to activities in the cenotes and adjacent ecosystems, especially the dry tropical forest [14]. This transition in tourism increases the threat to all underground systems because these new activities cause pollution by dissolved agents and solid wastes, as has been reported by several local organizations [33].

| System | Use |
|---|---|
| Crustacea | Little tourist activity |
| 27 Steps | Little tourist activity |
| Taj Maha | Frequent tourist activity |
| Casa Cenote | Frequent tourist activity |
| Vacaha | Frequent tourist activity |
| Calavera | Frequent tourist activity |
| Santa Cruz | Frequent tourist activity |
| Yum Ha | Little tourist activity |
| Rancho Regina | No tourist activity |
| Rancho San Felipe Nohoch | Frequent tourist activity |
| Aerolito | Little tourist activity |
| Xcan Ha | Frequent tourist activity |
| Tres Potrillos | No tourist activity |
| Cocodrilo | No tourist activity |
| Chankanaab | Tourist use only for observation |
| Chempita (Jade Caver) | Frequent tourist activity |
| Boca del Puma | Frequent tourist activity |
| Tres Bocas | Frequent tourist activity |
| Actun Jaleb | No tourist activity |
| Eden (jardin del EDEN) | Frequent tourist activity |
| Muevelo Rico | No tourist activity |
| Aktun Muknal | No tourist activity |
| Dos Arboles | No tourist activity |
| Odyssey | No tourist activity |
| Bang | No tourist activity |

**Table 4.**
*Tourist activity in cenotes.*

This region is experiencing a continual increase in tourism and mobility infrastructure [10] that together with the water demands evidently change the underground water conditions because exist the risk that if the pumping water increase the marine layer occupied these subterranean spaces and this work show the first photograph of these ecosystems under study which, possibly in one or two decades will be change. All these threats are present due to the economic benefits of tourist activity in the area. The different options to diminish the damage include environmental instruments and laws established by institutions such as the Federal Attorney for Environmental Protection (PROFEPA), Secretariat of Environment and Natural Resources (SEMARNAT), and National Commission of Natural Protected Areas Mexico (CONANP). Some proposed actions to reduce the harmful impacts of increased tourism include wastewater treatments, avoiding developing artificial greens like golf courses, stadiums, and parks, forbidding the injection of wastewater to deeper aquifers, and environmental education.

The discussed ecosystems are facing natural and anthropogenic impacts, highlighting their vulnerability. As such, monitoring their water conditions is highly important (**Tables 3** and **4**).

## 5. Conclusions

This chapter presented results of our study of groundwater conditions (temp, pH, oxygen dissolved, and salinity) in nineteen cenotes and underground rivers of the Mexican Caribbean corridor from Tulum to Puerto Morelos (Riviera Maya, Yucatan Peninsula) and six cenotes of Cozumel. We also profiled the predominant vegetation on the surface of this region, which is a seasonally dry tropical forest, to understand the components and functioning of these subterranean ecosystems to assess their vulnerability and identify their threats from human development (population growth, tourism development, mobility capacity). We identified three types of underground aquatic ecosystems: freshwater, brackish/marine water, and anchialine.

Rapid growth in tourism in the Riveria Maya and Cozumel, among other locations, is polluting and contaminating these regions' ecosystems and thus it is of great importance to monitor these ecosystems and the fauna that inhabit them to identify their capacity for resilience.

## Acknowledgements

The authors give thanks to Ana Vini for drawing the schematic view of the interaction between freshwater and marine layers. We are also grateful to the cave workers that helped us with our fieldwork: Germán Yañez, Guillermo Ruiz, Joey Pakes, Michel Vazquez, Fernando Paulin, and Amelia Weiss. This work is an outcome from projects supported by Quintana Roo Autonomous State University, the National Council of Science and Technology (CONACYT), and SEP-PRODEP.

## Conflict of interest

The authors declare no conflict of interest.

## Author details

Luis M. Mejía-Ortíz[1]*, Alejandro L. Collantes-Chávez-Costa[2], Cruz López-Contreras[3] and Oscar Frausto-Martínez[4]

1 Sustainable Development Division, Biospeleology and Carcinology Lab, Quintana Roo Autonomous State University, Mexico

2 Sustainable Development Division, Plant Ecology Lab, Quintana Roo Autonomous State University, Mexico

3 Tourism Department, Sustainable Development Division, Quintana Roo Autonomous State University, Mexico

4 Sustainable Development Division, Spatial Observatory Lab, Quintana Roo Autonomous State University, Mexico

*Address all correspondence to: luismejia@uqroo.edu.mx

## References


[1] Cejudo E, Ortega-Almazan PJ, Ortega-Camacho D, Acosta-González G. Hydrochemistry and water isotopes of a deep sinkhole in North Quintana Roo Mexico. Journal of South American Earth Sciences. 2022;**116**(103846):1-7

[2] Perry E, Velázquez OG, Marín L. The hydrogeochemistry of the karst aquifer system of the northern Yucatan Peninsula, Mexico. International Geology Review. 2002;**44**:191-221

[3] Bauer-Gottwein P, Gondwe B, Charvet G, Marín LE, Rebolledo-Vieyra M, Merediz-Alonso G. The Yucatán Peninsula karst aquifer, Mexico. Hydrogeology Journal. 2011;**19**(3):507-524

[4] Flores JS, Espejel I. Tipos de vegetación de la Península de Yucatán. In: Flores JS, editor. Etnoflora Yucatanense. Mérida Yucatán: Universidad Autónoma de Yucatán; 1994. p. 135

[5] Murphy PG, Lugo AE. Dry forest of Central America and the Caribbean Islands. In: Bullock SH, Mooney HA, Medina E, editors. Seasonally Dry Tropical Forest. New York: Cambridge University Press; 1995

[6] Mejía-Ortíz LM, Sprouse P, Tejeda-Mazariegos JC, Valladarez J, Frausto-Martínez O, Collantes-Chavez-Costa AL, et al. Tropical subterranean ecosystems in Mexico, Guatemala and Belize: A review of aquatic biodiversity and their ecological aspects. In: Natural History and Ecology of Mexico and Central America. Croacia: InTech Published; 2021

[7] Rodríguez-Castillo F, Frausto-Martínez O, Uhu-Yam WD, Colín-Olivares O. Morphometric characterisation of karst depressions: Coastal zone of the North-Eastern Yucatan Peninsula, Mexico. Ecosistemas y Recursos Agropecuarios. 2022;**9**(2) e3216:1-14

[8] Kambesis PN, Coke JG IV. The sac Actun system, Quintana Roo, Mexico. Boletín Geológico y Minero. 2016;**127**(1):177-192

[9] Back W, Hanshaw BB, Pyle TE, Plummer LN, Weidie AE. Geochemical significance of groundwater discharge and carbonate solution to the formation of Caleta Xel ha, Quintana Roo, Mexico. Water Resources Research. 1979;**15**:1521-1535

[10] Aguirre Barajas YM, Quezada Ramírez MF, Jáuregui DJ, A. Perfil sociodemográfico y laboral de los inmigrantes interestatales indígenas en Quintana Roo (México), 1990-2020. Notas de Población. 2022;**113**:193-216

[11] Castillo-Pavón O, Méndez-Ramírez JJ. Los desarrollos turísticos y sus efectos medioambientales en la Riviera Maya, 1980-2015. Quivera Revista de Estudios Territoriales. 2017;**19**(2):101-118

[12] El E-CH. origen del proyecto turístico Cancún, México: Una valoración de sus objetivos iniciales a 42 años de su nacimiento. Limina R. 2013;**11**(1):154-167

[13] Secretaría de Turismo. Rueda de prensa. [Gaceta]. 2022. Available from: https://sedeturqroo.gob.mx/ARCHIVOS/Gacetas/Gaceta-01-2022.pdf. [Accessed: May 15, 22]

[14] Consejo Civil Mexicano para la Silvicultura Sostenible 2022. Península

de Yucatán: las amenazas al territorio. CCMSSAC. 2022. Available from: http://ccmss.centrogeo.org.mx/cms/PYamenazasterritorio? [Accessed: June 13, 2022]

[15] Alvarez F, Iliffe TM, Benitez S, Brankovits D, Villalobos JL. New records of anchialine fauna from the Yucatan Peninsula Mexico. Checklist. 2015;**11**(1) 1505:1-10

[16] Mejía-Ortíz LM. Cave crustacean decapods from Mexico. In: Lachance N, editor. The Zoological Guide to Crustacea. New York, Nova Science Publisher; 2019. pp. 1-71

[17] Liévano-Beltrán LA, Simoes N. Update distribution of the mysid Antromysis cenotensis (Crustacea: Peracarida), a protected key species in Yucatan Peninsula cenotes. Diversity. 2021;**13**(154):1-17

[18] Pennington TD, Sarukhán J. Árboles tropicales de México: manual para la identificación de las principales especies. CdMx: UNAM; 2005

[19] Flores JS. Vegetación de cenotes, rejolladas, haltunes y cavernas. In: Durán- García R, Méndez-González M, editors. Biodiversidad y Desarrollo Humano en Yucatán. México DF. CICY, PPD-FMAM, CONABIO, SEDUMA, 2010. p. 43

[20] Yun Y, Wang H, Man B, Xiang X, Zhou J, Qiu X, et al. The relationship between pH and bacterial communities in a single karst ecosystems ad its implication for soil acidification. Frontiers in Microbiology. 1955;**2016**(7):1-14

[21] Cabrera Cano E, Sousa Sánchez M, Téllez Valdés O. Imágenes de la Flora Quintanarroense. Chetumal, Centro de investigaciones de Quintana Roo A.C. (CIQRO). 1982. p. 224

[22] Leopold AS. Vegetation zones of Mexico. Ecology. 1950;**31**(4):507-518

[23] Rzedowski J. Vegetación de México. 1ra. Edición digital, Comisión Nacional para el Conocimiento y Uso de la Biodiversidad. México: Cd; 2006

[24] Miranda F, Hernández EX. Los tipos de vegetación de México y su clasificación. Boletín de la Sociedad Botánica de México. 1963;**28**:29-178

[25] Köppen W. Climatología: Con un estudio de los climas de la tierra. México DF: Fondo de Cultura Económica; 1948. p. 479

[26] Durán García R, Contreras GG. Distribución espacial de la vegetación. In: Durán-García R, Méndez-González M, editors. Biodiversidad y Desarrollo 61 Humano en Yucatán. México D. F. CICY, PPD-FMAM, CONABIO, SEDUMA. 2011. pp. 131-135

[27] Enseñat F. Impactos y manejo del visitante en cenotes del caribe mexicano. In: Bautista F (Compilador), Los territorios karsticos de la península de Yucatán: caracterización, manejo y riesgos. Cd de México. Asociación Mexicana de Estudios sobre el Karst (AMEK). 2021. pp. 124-132

[28] Casas-Beltrán DA, Gallaher CM, Hernandez Yac E, Febles Moreno K, Voglesonger K, LealBautista RM, et al. Seaweed invasion! Temporal changes in beach conditions Lead to increasing cenote usage and contamination in the Riviera Maya. Sustainability. 2020;**12**(6):2474

[29] Aguilar Y. La importancia cultural, natural, ecológica y económica del anillo de cenotes: el caso de Homún. In: Bautista F (Compilador), Los territorios karsticos de la península de Yucatán: caracterización, manejo y riesgos. Cd. De

México. Asociación Mexicana de Estudios sobre el Karst (AMEK). 2021. pp. 161-171

[30] Cervantes-Martínez A, Gutiérrez-Aguirre MA. Copépodos en cenotes de Quintana Roo. ECOfronteras. 2017;**21**(59):7-7

[31] Ruíz-Ramírez JD, López Montejo GD, Cabrera GRJ. Manglares, desarrollo turístico y cambio climático en Playa del Carmen, corazón del caribe mexicano. Cuadernos de biodiversidad. 2018;**55**:28-40

[32] Mejía-Ortíz LM, Culver DC PT, Sprouse P. The blurred line between photic and aphotic environments: A large Mexican cave with almost no dark zone. International Journal of Speleology. 2018;**47**(1):1-12

[33] Casas-Beltrán DA, Hernández-Pedraza M, Alvarado-Flores J. Estimation of the discharge of sunscreens in aquatic environments of the Mexican Caribbean. Environments. 2020;**7**(2):1-11

Chapter 7

# Monitoring of Temporary Ponds as Indicators of Environmental Quality

*Claudia Campanale, Vito Felice Uricchio and Carmine Massarelli*

## Abstract

Temporary ponds represent a specific type of ecosystem extensively widespread worldwide. They are better known as copular pools, ephemeral waters, karst sinkholes, seasonal wetlands, and vernal pools. Among these, Mediterranean Temporary Ponds (MTPs) represent a priority habitat according to the Natura 2000 network of the European Union. Their main characteristic is represented by their depth of only a few centimeters and lack of communication with permanent water bodies. MTPs habitats are vulnerable to human activities, especially agriculture, and they are considered priority habitats to safeguard. Threats affecting this habitat are various and many and depend on specific site conditions, including intensive agriculture, tree planting, abandonment of traditional land use, and excessive grazing. In the present manuscript, we report the results of monitoring activity of some of these sites in Southern Italy aimed at understanding the ecological status of these ephemeral ecosystems with a specially developed methodology based on data integration.

**Keywords:** temporary ponds, microplastic pollution, pesticides, data integration, GIS

## 1. Introduction

Temporary ponds are a common natural habitat, abundantly widespread in all biogeographical regions [1]. Other ecological studies indicate that temporary ponds are habitats of some biological importance because they can host a considerable number of rare and endemic species [2–4]. These habitat types are currently highly threatened. Almost all temporary ponds are shallow, and most can be easily destroyed by drainage works for agricultural or urban development purposes [5]. Their small water volumes influence their high susceptiblity to pollution [6, 7], including emerging pollutants such as microplastics (MPs) and pesticides [8, 9]. To these threats is added that deriving from a lack of awareness: even if located within protected areas, temporary ponds have not always been evaluated by professionals in the same way as other freshwater ecosystems better known as lakes, rivers, and permanent ponds. Consequently, the conservation of the temporary ponds has never been to the attention of the administrators, and without any criteria, they have been destroyed over time for various purposes [2].

Among the threatening factors for these environments, we must also consider the ongoing climate changes on a global scale: it is probable that temporary ponds, with their delicately balanced hydrological regimes, are susceptible particularly to these changes and over a few years they could significantly reduce [10]. Another aspect of difficulty in providing correct information to the administrators of the territory to combine development needs with conservation criteria for these habitats is the lack of information.

Studies on temporary ponds are at least 50 years late compared to those of better-known water bodies, so this article aims to propose a multidisciplinary methodology to carry out the monitoring of temporary ponds to identify the pressures that insist on the hydrographic basin and sub-basin of interest. Temporary ponds (**Figure 1**) can be defined simply, as "lentic water bodies with a recurrent dry phase" [1]. This description includes a large range of water bodies, including tiny puddles that can hold water for only a few days and water bodies subjected to dryness after a few weeks or years. Like other freshwater habitats, temporary ponds are also changeable: as the duration of the hydroperiod increases, they turn into semi-permanent ponds and dry out only in years of drought.

A distinctive feature of temporary ponds is that they can form almost anywhere, it is only necessary for water to accumulate in a depression in the ground and silt to prevent water from infiltrating underground quickly. Their ease of formation and, at times, their persistence over time means that they are present almost everywhere in different types of ecosystems, both of natural and non-natural origin. They can form as a result of small barriers created by the fall of trees as in depressions caused by man: vehicle furrows, quarry bottoms, etc.

Among the temporary ponds, the ponds located in the Mediterranean region, so-called Mediterranean Temporary Ponds (MTPs) [11] are classified as priority habitats according to the Natura 2000 network of the European Union (Natura code 3170, Habitats Directive 92/43/EEC). They are situated in many Mediterranean countries. They are categorized as "priority" due to their elements with a unique and important meaning for one or more living species. They include a peculiar flora composition,

**Figure 1.**
*Some temporary ponds in the Puglia region (South Italy).*

succession stage, and/or structural factor. Since they are humid depressions, periodically subject to temporary seasonal submersions, they host plant communities of great richness and originality even if not very showy, they are rich from a floristic point of view with rare and exclusive species of these environments [12, 13]. Several temporary ponds in the Mediterranean region vary from small copular ponds 50 cm deep hollowed out in rocks to almost permanent lakes, sometimes covering more than several hectares. They present a consistent variability in size, shape, depth, biodiversity (flora and fauna), and time of flooding [14–18]. The increased urbanization and agriculture combined with climate change has led to the extinction of a vast number of temporary ponds in the Mediterranean region [18].

Temporary Mediterranean ponds exhibit significant variability regarding the length of their hydrological period. They often form in karst areas and are in equilibrium with the aquifer, which in some circumstances can lead to a rise in the water level. In natural karst environments, the permeability and slope and other geological properties determine how much water the duration of the hydroperiod and therefore the composition of the ecosystem generated by the temporary ponds.

As temporary ecosystems, MTPs represent an important and sensitive transition between aquatic and terrestrial environments and act as retention basins for macroplastics and MPs [8]. However, only a very few works to date have investigated the presence of MPs in temporary ponds and in particular in mountain karst ponds [19], rainwater retention ponds [20–22], and sport fishing ponds [23], or small water bodies [24].

Mediterranean temporary pond habitats are very susceptible to anthropic pressures [25], due to their particular physical and ecological features. Their value is frequently overlooked [26, 27] due to their small size and seasonality.

In many Mediterranean temporary ponds, human activities create great pressure. However, the high biodiversity that can be found has been preserved thanks to balances with human activities that have become less and less impacting [12, 27].

The development of agriculture and urbanization in the Mediterranean region influence the health status of numerous sites with temporary ponds [27, 28].

The increase in urbanization in the Mediterranean region has led, due to urban expansion and connecting infrastructures, to the extinction of numerous temporary ponds. Inadequate management practices, such as soil removal, drainage, overgrazing, and intensive agriculture, have seriously endangered these fragile ecosystems [29].

Intensive farming practices can lead to significant changes in the catchment area. Since a large part of the territory is ploughed, consequently, an increase in erosion is induced and the transported sediments are increasingly filling the small depressions of the ponds, thus modifying the hydroperiod.

The same phenomenon always has another consequence, namely an additional supply of nutrients to the ponds, thus contributing to their eutrophication and a decrease in environmental quality.

Furthermore, the pollution of temporary ponds is increased by intensive agricultural activities due to the extensive use of pesticides and fertilizers. Not infrequently, there is also the abandonment of domestic and industrial and polluting waste (as in the previous image) [30, 31]. The overload of nutrients reaching temporary ponds is a widespread danger in the Mediterranean-macroclimate territories. This occurs mainly due to the fertilizers run-off from neighboring areas [32].

Temporary ponds can be used to understand the quality of the environment of the basin concerned as being small areas in which pollutants transported in run-off water are deposited. So they can provide an assessment of the quality of the soil and water matrices of large areas by sampling at a single point and having some information about the places, also available from open-source databases [33]. Following this methodology, it is possible to use temporary ponds as indicators of quality and possible environmental degradation.

## 2. Material and methods

### 2.1 Study area

The temporary ponds investigated are located within the area of the *Lama Balice* Regional Natural Park, a territory that still preserves, despite the various scattered phenomena of consolidated and ongoing use, a high landscape quality, attributable to the typical features of the agricultural landscape of the Puglia countryside. These ponds are formed due to heavy rains and are very ephemeral as the karst landscape accentuates the leaching of water into the subsoil.

In the *Lama Balice* Torrent Basin, lithological terms of the Cretaceous carbonate succession of the *Murgia* emerge on which deposits of the Plio-Pleistocene coverage rest; there are also alluvial deposits from the Holocene age located on the bottom of the main erosive furrows. Referring to the geological map of Italy in 1:100,000 scale [34] and the geological map of the *Murge* and *Salento* in 1:250,000 scale [35], in the area of the *Lama Balice* Torrent basin (**Figure 2**), the following lithostratigraphic units are recognized: Murge's limestones, Gravina's calcarenite, terraced marine deposits, and alluvial deposits [36]. The water inputs come from rain, surface run-off, snow melt, and also from groundwater. Outflows occur by infiltration, overflow, and evapotranspiration. If the groundwater level is higher than the bottom of the pond, the pond water will tend to replenish the aquifer.

These environments, considering their proximity to an urban area, can play an essential role in maintaining and strengthening the link between human populations, wild flora, and fauna (mainly local and migratory birds) and can also be considered:

- A stepping-stone that is a fragment of natural habitat surrounded by an unsuitable landscape, which acts as a rest area, refuge, and dispersal for various species;
- A new biodiversity hotspot;
- A playful-educational and esthetic-recreational type tool for schools and cultural associations.

The protection of this pond is a great opportunity for the two districts. Its small size makes it easier to manage and protect and its perception by the public.

### 2.2 Water matrix sampling and chemical analyses

Three sites classified as MTPs habitats located in Southern Italy (Puglia region) and called MTP n.1, MTP n.2, and MTP n.3 were investigated to assess their role in the accumulation of Plant Protection Products (PPPs). Moreover, several environmental parameters were monitored on a seasonal basis at each site, including water depth,

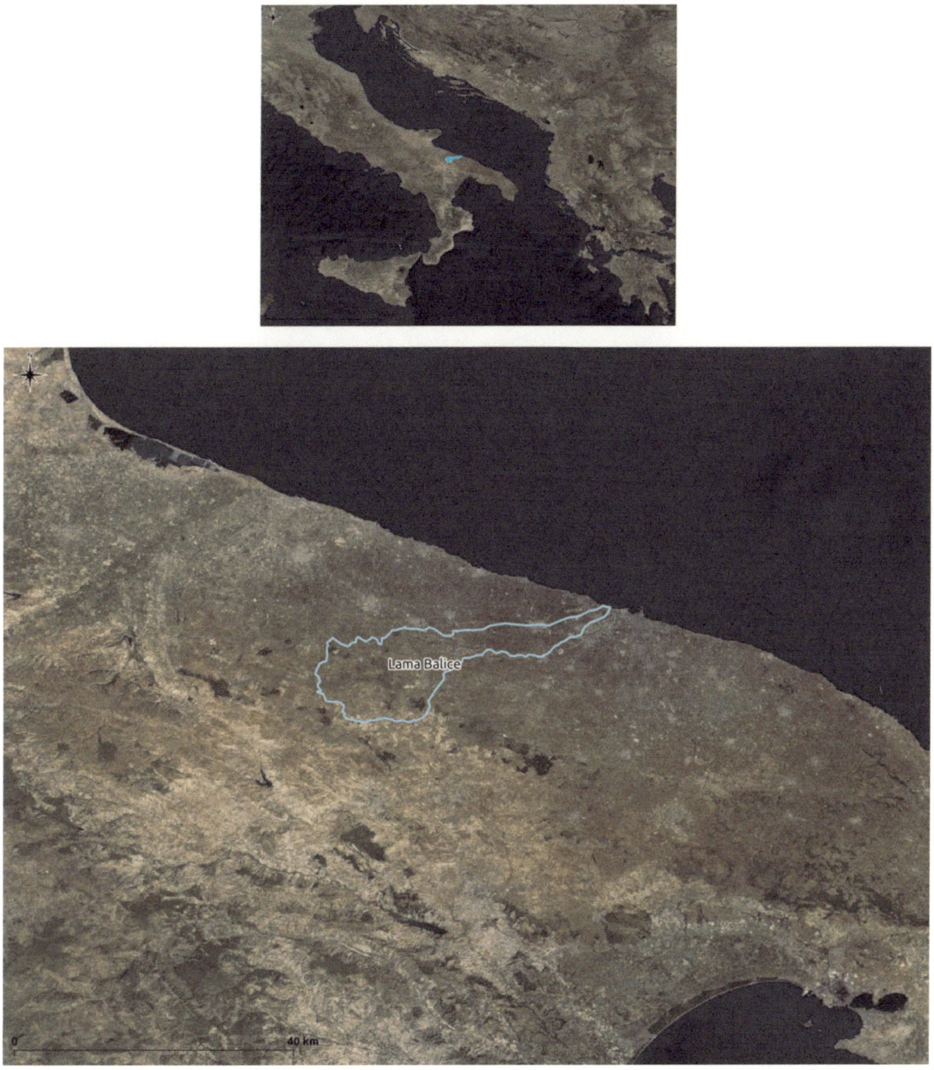


**Figure 2.**
*The* Lama Balice *Torrent basin (Base map from Bing Maps).*

temperature (T), pH, dissolved oxygen (D.O.), redox potential (ORP), salinity, electrical conductivity, total dissolved solids (TDSs), chemical oxygen demand (COD), nutrients (phosphates, nitrates, nitrites, and ammonium), and alkalinity.

Water samples composed of, at least three replicates for each site, were collected for a minimum of 3L of water volume for each sample and preserved in dark glass jars stored at 4°C until analyses [37].

Significant ions and nutrients were quantified by ion chromatography. Otherwise, the environmental parameters were measured *in situ* by a field multiparameter probe from HANNA Instruments.

The analysis of more than one hundred PPP residues was performed through Ultra High-Pressure Liquid Chromatography-Tandem Mass Spectrometry (UHPLC-MS/MS), mainly based on multiresidual methods.

Following is reported the list of the PPPs investigated (**Table 1**).

| n° | Substance | LOQ (µg/L) |
|---|---|---|
| 1 | 2.4-D | 0.005 |
| 2 | Acetamiprid | 0.005 |
| 3 | Acido Gibberellico | 0.025 |
| 7 | Azinfos etile | 0.005 |
| 8 | Azinfos metile | 0.005 |
| 9 | Azoxystrobin | 0.005 |
| 10 | Bensulfuron-methyl | 0.005 |
| 11 | Bentazone | 0.005 |
| 12 | Boscalid | 0.005 |
| 13 | Bromoxinil | 0.005 |
| 14 | Chlorantraniliprole | 0.005 |
| 15 | Chloridazon | 0.005 |
| 16 | Chlorotoluron | 0.005 |
| 17 | Chlotianidin | 0.005 |
| 18 | Cicloxidim | 0.005 |
| 19 | Cimoxanil | 0.005 |
| 20 | Clodinafop-Propargyl | 0.005 |
| 21 | Clopyralid | 0.005 |
| 22 | Cyazofamid | 0.005 |
| 23 | Cypermethrin | 0.025 |
| 24 | Cyprodinil | 0.005 |
| 25 | Dazomet | 0.025 |
| 26 | Deltamethrin | 0.005 |
| 27 | Demeton | 0.005 |
| 28 | Diclorvos | 0.005 |
| 29 | Difenoconazol | 0.005 |
| 30 | Dimethenamid | 0.005 |
| 31 | Dimethomorph | 0.005 |
| 32 | Ditianon | 0.025 |
| 33 | Diuron | 0.005 |
| 34 | Dodina | 0.005 |
| 35 | Ethofumesate | 0.005 |
| 36 | Etofenprox | 0.005 |
| 37 | Etoprofos | 0.005 |
| 38 | Fenamidone | 0.005 |
| 39 | Fenhexamid | 0.005 |
| 40 | Fenpirazamina | 0.005 |
| 41 | Fention | 0.01 |

| n° | Substance | LOQ (μg/L) |
|---|---|---|
| 42 | Flonicamid | 0.005 |
| 43 | Fluazifop-P-Butile | 0.005 |
| 44 | Fludioxonil | 0.005 |
| 45 | Flufenacet | 0.005 |
| 46 | Fluopicolide | 0.005 |
| 47 | Fluopyram | 0.005 |
| 48 | Fluroxypir | 0.01 |
| 49 | Formetanato | 0.005 |
| 50 | Hexythiazox | 0.005 |
| 51 | Imidacloprid | 0.005 |
| 52 | Ioxynil | 0.005 |
| 53 | Iprovalicarb | 0.005 |
| 54 | Isoproturon | 0.005 |
| 55 | Lenacil | 0.005 |
| 56 | Malation | 0.005 |
| 57 | Mandipropamid | 0.005 |
| 58 | Mcpa | 0.005 |
| 59 | Mecoprop (MCPP) | 0.005 |
| 60 | Mesosulfuron-Metil | 0.005 |
| 61 | Metalaxyl-M | 0.005 |
| 62 | Metamidofos | 0.005 |
| 63 | Metamitron | 0.005 |
| 64 | Metazaclor | 0.005 |
| 65 | Methiocarb | 0.005 |
| 66 | Metomil | 0.005 |
| 67 | Metossifenozide | 0.005 |
| 68 | Metrafenone | 0.005 |
| 69 | Mevinfos | 0.005 |
| 70 | Myclobutanil | 0.005 |
| 71 | Oxadiazon | 0.005 |
| 72 | Oxamil | 0.005 |
| 73 | Penconazol | 0.005 |
| 74 | Phenmedipham | 0.025 |
| 75 | Phosmet | 0.005 |
| 76 | Picoxystrobin | 0.005 |
| 77 | Pinoxaden | 0.005 |
| 78 | Prochloraz | 0.005 |
| 79 | Propamocarb | 0.005 |

| n° | Substance | LOQ (μg/L) |
|---|---|---|
| 80 | Propiconazolo | 0.005 |
| 81 | Propizamide | 0.005 |
| 82 | Pyraclostrobin | 0.005 |
| 83 | Pyrimethanil | 0.005 |
| 84 | Quinoxyfen | 0.005 |
| 85 | Sedaxane | 0.005 |
| 86 | Spinosad | 0.025 |
| 87 | Spiroxamine | 0.005 |
| 88 | Tebufenpyrad | 0.005 |
| 89 | Terbutilazina.desetil-(metabolita) | 0.005 |
| 90 | Terbutrina | 0.025 |
| 91 | Tetraconazole | 0.005 |
| 92 | Thiamethoxam | 0.005 |
| 93 | Thiophanate-Methyl | 0.005 |
| 94 | Tiacloprid | 0.005 |
| 95 | Tolclofos-Methyl | 0.025 |
| 96 | Triallate | 0.005 |
| 97 | Tribenuron Metile | 0.005 |
| 98 | Triciclazolo | 0.005 |
| 99 | Zoxamide | 0.005 |

**Table 1.**
*PPPs analyzed by multiresidual method and related limits of quantification (LOQ).*

Regarding glyphosate, AMPA, and glufosinate, a test aliquot from each water sample was treated following the derivatization procedure described by [38].

The instrumental analysis was conducted with a triple quadrupole mass-spectrometer system (TSQ Altis, Thermo Scientific, Massachusetts, USA) equipped with an Electro Spray Ionisation (ESI) source and coupled to a Vanquish Horizon UHPLC System (Thermo Scientific, Massachusetts, USA).

## 2.3 Insect monitoring

Reference studies on insect populations are increasingly relevant and necessary in the middle of the acceleration of concern for current trends of insects. The need to monitor insects is increasingly emerging as considered in continuous decline and above all because they can provide an ecological response based on many aspects such as occurrence and distribution, phenology, abundance and biomass, diversity and composition of the species [39]. Visual investigations are commonly used to document the abundance and diversity of insects that can be easily identified on the field, often with the help of binoculars and close focus networks. These investigations generally involve researchers who document the presence of a species or count the total number of individuals of each species observed during a standardized investigation. The most frequently used methods include (1) transepts, (2) counts for points, and (3) counts for areas. We specify that visual investigations are commonly used to document the abundance and diversity of insects that can be easily identified on the field, often with the help of binoculars and nets with a closed focus. These investigations generally involve researchers who document the presence of a species or count the total number of individuals of each species observed during a standardized investigation. For this type of monitoring, 1 km transepts were considered with the high sun on a windless and rainless day.

### 2.4 GIS procedure

To identify a functional operational flow, the areas of interest that act as a hydrographic basin for each temporary pond identified in the area along the main branch of the *Lama Balice* were first identified. This procedure was implemented in a Geographical Information System (GIS) software by processing the Digital Elevation Model (DEM) of the Puglia region with the module watershed basins in the open-source software SAGA GIS [40, 41].

In addition, for each sub-basin, again through a GIS-based procedure, the main crops present are identified and sorted from largest to smallest by extension in hectares. This procedure makes it possible to correlate any identified pesticide residues with the prevailing crops [30].

Finally, always through this methodology, it is possible to identify the areas of the territory where other temporary ponds could potentially form.

### 2.5 Landscape metrics

Again through GIS and supported by the use of recent high-definition maps, such as those achievable through a drone [42], it is possible to calculate the following metrics relating to the anthropogenic activities associated with disturbance activities and which can contribute in various ways to the evaluation of the quality status of the water matrix of the ponds. These metrics consist of mapping quarries, paths, and land cover changes for each sub-basin to obtain information with high added value with the analysis of the distance matrix and concentration maps. Specifically, the assessments are carried out through the calculation of anthropic pressure indicators such as the degree of fragmentation of the biotope produced by the road network (IND1PA), constriction of the biotope (IND2PA), diffusion of anthropic disturbance (IND3PA) [43], and as foreseen in the Manuals for habitat monitoring by ISPRA [44] get this information:

1. Nearness to environmental detractors, such as:

    - mining and landfills;
    - potentially contaminated areas;
    - contaminated areas;
    - vulnerable areas from nitrates of agricultural origin and vulnerable areas from plant protection products (also useful for measuring the pressure on the habitat due to agricultural activities);

- areas at risk of desertification;
- areas with strong tourist pressure;
- proximity to treatment plants and pipelines for urban wastewater;
- pollution of surface waters (if applicable) measured as pollution by N, P, BOD due to the drainage basin;

2. Pressure on habitat due to agricultural activities is measured as:
   - nearness to agricultural activities and to areas vulnerable to nitrates of agricultural origin and vulnerable areas to plant protection products;
   - assessment of the degree of habitat fragmentation based on Corine Land Cover [45];
3. Proximity to the road network is measured as the distance of the habitat from the nearest road segment;
4. Habitat consumption measured in % of the habitat occupied by anthropogenic artifacts;
5. Proximity to an airport facility is measured in % of the habitat within 5 km of an airport.

### 2.6 Field surveys

During the field surveys, it is advisable to note down what may be additional disturbing factors of the habitat, with an intensity scale that will allow identifying the factors that most negatively affect the habitat: for example, waste abandonment practices (it is suggested to count only areas in which the abandonments cover an area greater than 30 square meters), massive spreading of livestock effluents, and any other aspect that could interfere with the naturalness of an area.

### 2.7 Data integration

Given the different data sources, it was necessary to think about a data integration methodology. It is often necessary to compare different types of data, and therefore it is necessary to make them comparable through the creation of a single unified dataset. The resulting dataset differs from a simply combined superset in that the points in the new dataset contain information that is new to the data that originated it.

To this end, a form for field surveys was created by integrating other aspects required by the Italian Environmental Agency (ISPRA) [46] with normalized scales to be able to compare the collected data. The purpose of this action is precisely to standardize a method for fine-tuning the estimate of anthropogenic activities on habitats and ecosystems capable of combining multiple cognitive needs.

## 3. Results and discussion

### 3.1 Results of chemical investigations

Following (**Tables 2** and **3**) the results of the chemical investigations carried out in the three MTPs identified are reported.

As far as the environmental parameters and the concentration of the nutrient observed, the MTP n.3 resulted in the richest of the investigated elements among the three sites selected. The proximity to the marine environment is demonstrated by the higher levels of conductivity and salinity encountered compared to the other two internal sites. Noteworthy is also the water temperature of 24°C revealed in MTP n.3 very higher compared to the other two sites, and it is also greater concerning the mean seasonal values.

| Parameter | *Unit of measure* | MTP n.1 | MTP n.2 | MTP n.3 |
|---|---|---|---|---|
| pH | | 8.28 | 7.2 | 8.3 |
| D.O. | % | 120.55 | 47 | 281.2 |
| D.O. | mg/L | 9.6 | 3.77 | 19.4 |
| ORP | mV | 139.4 | 158.2 | 102 |
| CONDUCTIVITY | ms/cm | 09.46 | 11.28 | 23.07 |
| TDS | ppm | 9.74 | 10.14 | 11.65 |
| SALINITY | PSU | 08.63 | 9.16 | 14 |
| T | °C | 19.63 | 18.47 | 24.5 |
| PHOSPHATES | mg/L | 0.32 | 0.13 | 0.15 |
| NITRATES | mg/L | < 1 | < 1 | < 1 |
| AMMONIUM | mg/L | <0.02 | <0.02 | 0.258 |
| NITRITES | mg/L | <0.02 | <0.02 | 0.0329 |
| COD | mg/L | 31.2 | 37.4 | 113.4 |

**Table 2.**
*Results of the analyses of the environmental parameters and nutrients investigated in the three identified study sites.*

| PPP name | Biocide action | CAS | MF | MW | LOQ (μg/L) | Concentration (μg/L) | | |
|---|---|---|---|---|---|---|---|---|
| | | | | | | MTP n.1 | MTP n.2 | MTP n.3 |
| Acetamiprid | Insecticide | 135410-20-7 | $C_{10}H_{11}ClN_4$ | 222.67 | 0.005 | < LOQ | < LOQ | < LOQ |
| AMPA | Herbicide | 1066-51-9 | $CH_6NO_3P$ | 111.0 | 0.025 | 0.075 | < LOQ | 0.085 |
| Bentazone | Herbicide | 25057-89-0 | $C_{10}H_{12}N_2O_3S$ | 240.28 | 0.005 | < LOQ | < LOQ | 0.006 |
| Carbendazim | Fungicide | 10605-21-7 | $C_9H_9N_3O_2$ | 191.19 | 0.005 | < LOQ | < LOQ | 0.05 |
| Glyphosate | Herbicide | 1071-83-6 | $C_3H_8NO_5P$ | 169.1 | 0.025 | < LOQ | < LOQ | 0.05 |
| Imidacloprid | Insecticide | 138261-41-3 | $C_9H_{10}ClN_5O_2$ | 255.66 | 0.005 | < LOQ | < LOQ | 0.09 |
| Metalaxyl | Fungicide | 57837-19-1 | $C_{15}H_{21}NO_4$ | 279.33 | 0.005 | < LOQ | < LOQ | 0.006 |

*MF: molecular formula; MW: molecular weight; and LOQ: limit of quantification.*

**Table 3.**
*Results of the PPPs investigated in the three identified study sites.*

Regarding the residues of the PPP investigated, about 7 % of the substances analyzed were revealed in a quantifiable amount. Among the substances positively quantified, two of these were insecticides (Acetamiprid and Imidacloprid), two fungicides (Carbendazim and Metalaxyl), and three herbicides (Bentazone, Glyphosate, and AMPA). In the MTP n.3, all these pesticides were found to be above the limit of quantification and with higher concentrations. On the contrary, in MTP n.2, just two of these were positively quantified and in MTP n.1 six of seven. At no site were any exceeding of the regulatory threshold values found.

### 3.2 Results of the GIS procedure

The results of the applied GIS procedure are reported below. In **Figure 3**, it is possible to observe the subdivision of the main basin into subbasins and the actual coincidence, with margins of error of a few meters (absolutely acceptable given the presence of morphological aspects that can alter the natural conformation of the territory such as roads and small bridges) of the temporary ponds with their closing sections. It is interesting to note that other temporary ponds could potentially form in areas with other closed sections.

The identification of sub-basins and cultivation practices for each of them is very important as it allows to identify the potential origin of the sources of contamination that have the greatest impact on these delicate ecosystems. The proposed monitoring sheet is filled out for each of them.

### 3.3 Results of insect monitoring

The results relating to the number of identified insects, for each monitored site, are reported directly in **Table 4**.

**Figure 3.**
*Basins and sub-basins identified.*

| | Monitored sites n. | %x10 Land Use of the sub-Bacin | | | n. of illegal landfills for km$^2$ | n. of water discharges for km$^2$ | n. sites with sparks of zootechnical effluents for km$^2$ | others for km$^2$ | Historical information, for example reclaimed sites for km2 | n. asphalted roads that intersect a transept | n. diseases identified on plants | Insects | | sum of the voices and subtraction of the number of useful insects | Mean | Number of pesticides found | Number of contaminants (metals) found | TOTAL |
|---|---|---|---|---|---|---|---|---|---|---|---|---|---|---|---|---|---|---|
| | | | | | max 10 | max 10 | max 10 | max 10 | | | max 10 | | | | | | | |
| | | Agricultural | Pasture | Urban | | | | | | | | n. exclusive species | n. opportunistic species | | | | | |
| Sub-Basin n. 1 | 1 | 9 | 1 | 0 | 1 | 0 | 0 | 0 | 1 | 0 | 2 | 5 | 5 | 4 | 3.00 | 1 | 0 | 0.25 |
| | 2 | 8 | 2 | 0 | 0 | 0 | 0 | 0 | 0 | 0 | 1 | 4 | 5 | 2 | | | | |
| Sub-Basin n. 2 | 1 | 9 | 1 | 0 | 0 | 0 | 1 | 0 | 0 | 1 | 0 | 7 | 8 | 3 | 2.67 | 0 | 0 | 0.38 |
| | 2 | 9 | 1 | 0 | 0 | 0 | 0 | 0 | 0 | 1 | 0 | 6 | 8 | 3 | | | | |
| | 3 | 9 | 1 | 0 | 0 | 0 | 0 | 0 | 0 | 1 | 0 | 7 | 8 | 2 | | | | |
| Sub-Basin n. 3 | 1 | 8 | 1 | 2 | 2 | 0 | 1 | 0 | 2 | 2 | 3 | 2 | 6 | 14 | 11.00 | 5 | 3 | 0.05 |
| | 2 | 9 | 1 | 0 | 1 | 0 | 0 | 0 | 0 | 1 | 1 | 1 | 6 | 8 | | | | |
| Sub-Basin n. 4 | 1 | 7 | 1 | 2 | 3 | 0 | 1 | 0 | 2 | 1 | 3 | 1 | 7 | 16 | 14.50 | 5 | 3 | 0.04 |
| | 2 | 7 | 2 | 1 | 2 | 0 | 0 | 0 | 2 | 1 | 4 | 2 | 6 | 13 | | | | |

| | | | | | | | | | | | | | | | | | | |
|---|---|---|---|---|---|---|---|---|---|---|---|---|---|---|---|---|---|---|
| Sub-Basin n. 5 | 1 | 7 | 1 | 2 | 3 | 0 | 2 | 0 | 3 | 1 | 2 | 1 | 7 | 17 | 14.50 | 5 | 3 | 0.04 |
| | 2 | 7 | 2 | 1 | 2 | 0 | 1 | 0 | 2 | 1 | 2 | 2 | 6 | 12 | | | | |
| Sub-Basin n. 6 | 1 | 7 | 1 | 2 | 2 | 0 | 2 | 0 | 3 | 1 | 3 | 3 | 7 | 15 | 14.00 | 5 | 3 | 0.05 |
| | 2 | 7 | 2 | 1 | 2 | 0 | 1 | 0 | 2 | 1 | 3 | 2 | 6 | 13 | | | | |
| Sub-Basin n. 7 | 1 | 6 | 1 | 3 | 8 | 1 | 3 | 1 | 2 | 4 | 3 | 2 | 10 | 30 | 30.67 | 5 | 3 | 0.03 |
| | 2 | 6 | 1 | 3 | 6 | 2 | 2 | 2 | 1 | 3 | 4 | 1 | 9 | 28 | | | | |
| | 3 | 5 | 2 | 3 | 6 | 0 | 3 | 2 | 4 | 5 | 5 | 0 | 9 | 34 | | | | |

| Legend of the classification | |
|---|---|
| 0 | very bad |
| 0,1 | bad |
| 0,2 | scarse |
| 0,3 | medium |
| 0,4 | good |
| 0,5 | very good |

**Table 4.**
*Results achieved after applying the methodology for each monitored temporary pond.*

### 3.4 Data integration results

Below are the results achieved for each temporary pond monitored and the relative monitoring carried out at different points for each sub-basin. **Table 1** shows the data for a simple and immediate comparison.

In **Figure 4**, the resulting data integration values mapped on GIS are reported.

## 4. Discussion

The results show that along the path from upstream to downstream of the entire catchment area of the *Lama Balice*, n. 3 different situations: upstream, the quality state can be defined as intermediate and going downstream, it first becomes good and then worsens. This can be explained by the presence of a highly natural area between the mountain area where the city of *Bitonto* is located and the valley area where, in addition to the city of *Bari*, there are numerous anthropogenic activities, especially of industrial type.

Rather than evaluating the absolute value of the score obtained, this methodology is considered more appropriate for a relative evaluation of the state of the places. The colors that are assigned also represent a priority for intervention in certain areas and facilitate the understanding of dynamics in progress: for example, discontinuous colors suggest the presence of limited and site-specific changes.

The different inspections carried out and the results of the application of this methodology made it possible to identify some best practices to be implemented for the management of the embankments, such as:

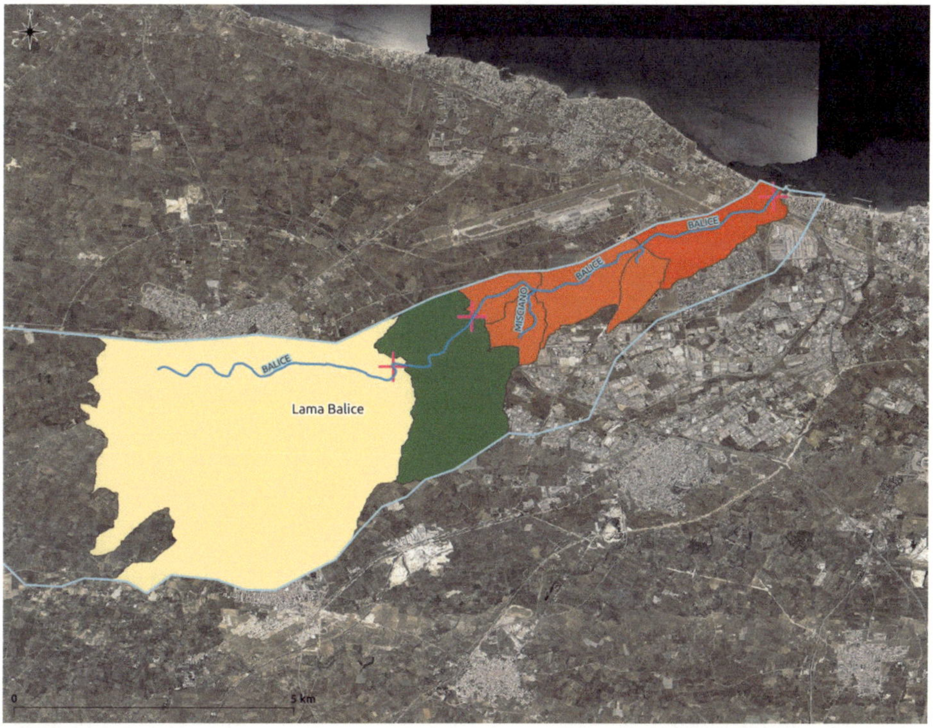


**Figure 4.**
*The classification of the state of sub-basins carried out through the monitoring of temporary ponds.*

- avoiding the use of herbicides to control the vegetation of the ditches and their embankments and also to maintain a high diversity of habitats along the banks of the canals;
- favor grazing along the edges of the ditches to the advantage of annual plants and some invertebrates;
- keep isolated trees and patches of shrubs to allow shading of large sections of the ditches;
- avoid planting new trees or hedges with species that are not typical of the ecosystem and keep existing plants low and manage shrub vegetation along the ditches to increase the presence of waterfowl and other animal species;
- maintain these ecosystems to create suitable habitats for invertebrates, a source of food for avifauna;

with the following application implications:

- review of the authorization processes (discharges and withdrawals);
- review of environmental impact assessment procedures so that they take into account these fragile ecosystems;
- implementation of the codes of good agricultural practice that fall within the field of measures aimed at achieving/maintaining the quality objectives of a water body according to the DQA.

The innovative aspects that can be introduced by adopting this methodology refer to the systematic use in operational investigation practices of methodologies based on "knowledge management" that allows the collection, evaluation, analysis, integration, and interpretation of all the information available regarding decision-making or investigative need, allowing to represent the interactions and evolutions. This technique favors the deepening of the theoretical bases of predictivity, through the more intrinsic analysis of the cases that led to the configuration of environmental pollution, the reconstruction of the relative model and the interpolation, for predictive purposes, of what and when it may occur. The development of the monitoring method, predictive analysis, the creation of concentration maps, and overall the creation of an information system for the management and use of the data collected and transmitted can be based on techniques of knowledge extraction through Machine methodologies. Learning and Data Mining (clustering, sequence clustering, decision trees, time series, and logistic regression) and Association Rules Discovery to identify information based on associative rules are able to describe an interesting relationship between different phenomena taking place in extremely complex environments and ephemeral [47].

## 5. Conclusions

It is most important to recognize the crucial ecological role of temporary ponds. The numerous threats interesting these habitats are gradually causing their extinction, especially in the Mediterranean region where, unfortunately, not all states must comply with European legislation because they are not included in the Economic Community. Therefore, the aforementioned environmental legislation does not apply to them. However, the conservation of ponds, regardless of laws and states of origin, is essential, and appropriate political and management measures should be taken immediately to prevent their disappearance.

The results of the present study show how temporary ponds can be used to improve the knowledge necessary for understanding the matrices' quality status in the River Basin of interest. The application of the proposed methodology made it possible to identify that in some areas, there is a greater risk on the water matrix and in others on the soil matrix based on a retro analysis based on the evaluation of pressures.

As far as water bodies are concerned, this type of monitoring not only allows the identification of a series of site-specific pressures but also an assessment of the state of the water matrix when monitoring operations are challenging to implement due to environmental conditions (e.g. dense vegetation or the presence of difficult accesses). In addition, considering these types of torrents, they are naturally created in a karst landscape, and therefore, their monitoring is carried out only on underground water bodies increasing costs for the opening through excavation and maintenance of wells. With this methodology, monitoring surface water bodies could be extended by increasing the cognitive picture of the quality of the water matrix with a minimum cost.

The integrated evaluation of the analysis of pressures and monitoring data can therefore be used to guide environmental control activities based on criteria of priority. This approach makes it possible to integrate the results of environmental controls, for example, both in the planning of monitoring and in the definition of protection measures.

## Acknowledgements

The authors thank Domenico Bellifemine for his valuable support in data acquisition campaigns.

## Author details

Claudia Campanale, Vito Felice Uricchio and Carmine Massarelli*
Water Research Institute of Italian National Research Council, Italy

*Address all correspondence to: carmine.massarelli@ba.irsa.cnr.it

## References

[1] Williams P, Biggs J, Fox G, Nicolet P, Whitfield M. History, origins and importance of temporary ponds. In: Freshwater Forum. Freshwater Biological Association (FBA): Ambleside. Vol. 17. 2001. pp. 7-15. ISSN 0961-4664; e-ISSN 2042-5244

[2] Bratton JH. Seasonal pools - an overlooked invertebrate habit. British Wildlife. 1990;**2**:22-29. DOI: 10.1016/0006-3207(92)90621-S

[3] Baskin Y. California's ephemeral vernal pools may be a good model for speciation. Bioscience. 1994;**44**:384-388. DOI: 10.2307/1312360

[4] King JL, Simovich MA, Brusca RC. Species richness, endemism and ecology of crustacean assemblages in northern California vernal pools. Hydrobiology. 1996;**3282**:328

[5] Serrano L, Serrano L. Influence of Groundwater Exploitation for Urban Water Supply on Temporary Ponds from the Doñana National Park (SW Spain). Journal of Environmental Management. 1996;**46**:229-238. DOI: 10.1006/JEMA.1996.0018

[6] Williams P. Lowland Pond Survey. London: DETR; 1996

[7] Massarelli C, Losacco D, Tumolo M, Campanale C, Uricchio VF, Costa-Pereira F, et al. Protection of water resources from agriculture pollution: An integrated methodological approach for the nitrates Directive 91–676-EEC implementation. Journal of Environmental Research and Public Health. 2021;**18**:13323

[8] Campanale C, Galafassi S, Savino I, Massarelli C, Ancona V, Volta P, et al. Microplastics pollution in the terrestrial environments: Poorly known diffuse sources and implications for plants. Science Total Environment. 2022;**805**: 150431. DOI: 10.1016/j.scitotenv.2021.150431

[9] Campanale C, Massarelli C, Bagnuolo G, Savino I, Uricchio VF. The Problem of Microplastics and Regulatory Strategies in Italy. Chem: Handbook Environment; 2019

[10] Acreman MC. The hydrology of the UK: A study of change. 2000

[11] Scheda tipo di habitat. Available from: http://vnr.unipg.it/habitat/cerca.do?formato=stampa&idSegnalazione=4. [Accessed: June 22, 2022]

[12] Grillas P, Gauthier P, Yavercovski N, Perennou C. Mediterranean temporary pools: Issues relating to conservation, functioning and management. Volume 1.; Station Biologique de la Tour Valat, 2004

[13] Ernandes P, Beccarisi L, Gigante D, Venanzoni R, Zuccarello V. Rare species of Mediterranean temporary pools in Apulia: New records and updating about distribution. Italian Botany. 2010;**42**:465-471

[14] Bagella S, Gascón S, Filigheddu R, Cogoni A, Boix D. Mediterranean Temporary Ponds: New challenges from a neglected habitat. Hydrobiologia. 2016; **782**:1-10

[15] Rhazi L, Grillas P, Rhazi M, Aznar JC. Ten-year dynamics of vegetation in a Mediterranean temporary pool in western Morocco. Hydrobiologia. 2009;**634**:185-194. DOI: 10.1007/s10750-009-9893-7

[16] Campos M, de Banyolas A, Patrick Grillas S, du Valat T,

Joaquín Fernández F. The European Commission (DG ENV B2) commissioned the Management of Natura 2000 habitats. 3170 *Mediterranean temporary ponds. 2000

[17] Dimitriou E, Karaouzas I, Skoulikidis N, Zacharias I. Assessing the environmental status of Mediterranean temporary ponds in Greece. Annales de Limnologie. 2006;**42**:33-41. DOI: 10.1051/limn/2006004

[18] Zacharias I, Zamparas M, Zacharias I, Zamparas ÁM. Mediterranean temporary ponds. Ecosystem. 2010;**19**:3827-3834. DOI: 10.1007/s10531-010-9933-7

[19] Iannella M, Console G, D'Alessandro P, Cerasoli F, Mantoni C, Ruggieri F, et al. Preliminary analysis of the diet of triturus carnifex and pollution in mountain Karst ponds in central apennines. Water (Switzerland). 2020; **12**:44. DOI: 10.3390/w12010044

[20] Liu F, Olesen KB, Borregaard AR, Vollertsen J. Microplastics in urban and highway stormwater retention ponds. Science Total Environment. 2019;**671**: 992-1000. DOI: 10.1016/j.scitotenv.2019.03.416

[21] Liu F, Vianello A, Vollertsen J. Retention of microplastics in sediments of urban and highway stormwater retention ponds. Environmental Pollution. 2019;**255**:113335. DOI: 10.1016/j.envpol.2019.113335

[22] Olesen KB, Stephansen DA, van Alst N, Vollertsen J. Microplastics in a stormwater pond. Water (Switzerland). 2019;**11**:1466. DOI: 10.3390/w11071466

[23] Campanale C, Massarelli C, Savino I, Locaputo V, Uricchio VF. A detailed review study on potential effects of microplastics and additives of concern on human health. International Journal of Environmental Research and Public Health. 2020;**17**(4):1212. DOI: 10.3390/ijerph17041212

[24] Hu L, He D, Shi H. Microplastics in Inland Small Waterbodies. Handbook Environmental Chemistry. 2020;**95**: 93-110. DOI: 10.1007/698_2019_445

[25] Massarelli C. Fast detection of significantly transformed areas due to illegal waste burial with a procedure applicable to Landsat images. International Journal of Remote Sensing. 2018;**39**:754-769

[26] Schwartz SS, Jenkins DG. Temporary aquatic habitats: Constraints and opportunities. Aquatic Ecology. 2000;**341**:3-8

[27] Beja P, Alcazar R. Conservation of Mediterranean temporary ponds under agricultural intensification: An evaluation using amphibians. Biological Conservation. 2003;**114**:317-326. DOI: 10.1016/S0006-3207(03)00051-X

[28] Brendonck L, Williams WD. Biodiversity in wetlands of dry regions (drylands). 2000

[29] Aponte C, Kazakis G, Ghosn D, Papanastasis VP. Characteristics of the soil seed bank in Mediterranean temporary ponds and its role in ecosystem dynamics. Wetlands Ecology and Management. 2010;**18**:243-253. DOI: 10.1007/S11273-009-9163-5/FIGURES/3

[30] Massarelli C, Ancona V, Galeone C, Uricchio VF. Methodology for the implementation of monitoring plans with different spatial and temporal scales of plant protection products residues in water bodies based on site-specific environmental pressures assessments. Human and Ecological Risk Assessment:

An International Journal. 2019;**26**(5): 1341-1358. DOI: 10.1080/10807039.2019.1578945

[31] Lahr J, Diallo AO, Gadji B, Diouf PS, Bedaux JJM, Badji A, et al. Ecological effects of experimental insecticide applications on invertebrates in sahelian temporary ponds. Environmental Toxicology and Chemistry. 2000;**19**: 1278-1289. DOI: 10.1002/etc.5620190509

[32] Losacco D, Ancona V, De Paola D, Tumolo M, Massarelli C, Gatto A, et al. Development of ecological strategies for the recovery of the main nitrogen agricultural pollutants: A review on environmental sustainability in agroecosystems. Sustainable. 2021;**13**:7163

[33] OpenStreetMap. Available from: https://www.openstreetmap.org/copyright. [Accessed: June 22, 2022]

[34] Servizio Geologico D'italia Geologica d'Italia, in scala 1:100.000. F° 177 «Bari» and F° 189 «Altamura» . Libr. di Stato 1966

[35] Ciaranfi N, Pieri P, Ricchetti G. Note alla carta geologica delle Murge e del Salento (Puglia centromeridionale). Mem. Soc. Geol. It. 1988;**41**:449-460

[36] Massarelli C, Campanale C, Uricchio VF. Ground penetrating radar as a functional tool to outline the presence of buried waste: A Case Study in South Italy. Sustainable. 2021;**13**:3805

[37] Campanale C, Massarelli C, Losacco D, Bisaccia D, Triozzi M, Uricchio VF. The monitoring of pesticides in water matrices and the analytical criticalities: A review. TrAC Trends in Analytical Chemistry. 2021; **144**:116423. DOI: 10.1016/J.TRAC.2021.116423

[38] Campanale C, Triozzi M, Massarelli C, Uricchio VF. Development of a UHPLC-MS/MS method to enhance the detection of Glyphosate, AMPA and Glufosinate at sub-microgram / L levels in water samples. Journal of Chromatography A. 1672;**2022**:463028. DOI: 10.1016/J.CHROMA.2022.463028

[39] Montgomery GA, Belitz MW, Guralnick RP, Tingley MW. Standards and best practices for monitoring and benchmarking insects. Frontiers in Ecology and Evolution. 2021;**8**:513. DOI: 10.3389/FEVO.2020.579193/BIBTEX

[40] Conrad O, Bechtel B, Bock M, Dietrich H, Fischer E, Gerlitz L, et al. System for Automated Geoscientific Analyses (SAGA) v. 2.1.4 System for Automated Geoscientific Analyses (SAGA) v. 2.1.4 System for Automated Geoscientific Analyses (SAGA) v. 2.1.4. Geoscience Modelling Developmental Discussion. 2015;**8**:2271-2312. DOI: 10.5194/gmdd-8-2271-2015

[41] Module Watershed Basins / SAGA-GIS Module Library Documentation (v2.1.3). Available from: https://saga-gis.sourceforge.io/saga_tool_doc/2.1.3/ta_channels_1.html [Accessed: July 4, 2022]

[42] Massarelli C, Matarrese R, Uricchio VF, Muolo MR, Laterza M, Ernesto L. Detection of asbestos-containing materials in agro-ecosystem by the use of airborne hyperspectral CASI-1500 sensor including the limited use of two UAVs equipped with RGB cameras. International Journal of Remote Sensing. 2017;**38**:2135-2149. DOI: 10.1080/01431161.2016.1226528

[43] VV.AA. The Nature Map of the Puglia Region. Rome; 2014

[44] VV.AA. Manuals for Monitoring of Species and Habitats of Community

Interest (Directive 92/43/EEC) in Italy: Habitat. Rome; 2016

[45] Riitters K, Wickham J, O'Neill R, Jones B, Smith E. Global-scale patterns of forest fragmentation. Conservation Ecological Publication. 2000

[46] VV.AA. Guidelines for the Analysis of Pressure According to Directive 2000/60/EC. Italian National System for Environmental Protection. Rome. Document number 26/2018. pp. 1-105

[47] Massarelli C, Campanale C, Uricchio VF. Artificial intelligence and water cycle management. IoT Applied Computer. 2021. DOI: 10.5772/INTECHOPEN.97385